W9-ARV-370

# THE CLIMATIC ATLAS OF MICHIGAN

Western Michigan University Department of Geography CartoGraphics Laboratory Staff:

Christopher T. Forth
Douglas Kyle
Mary Dillworth
Tom Haney

# THE CLIMATIC ATLAS OF MICHIGAN

**Val L. Eichenlaub**
Professor of Geography, Western Michigan University

**Jay R. Harman**
Professor of Geography, Michigan State University

**Fred V. Nurnberger**
State Climatologist, Michigan Department of Agriculture, Climatology Program,
Adjunct Professor of Geography, Michigan State University

CARTOGRAPHIC DIRECTOR
**Hans J. Stolle**
Associate Professor of Geography, Western Michigan University

**The University of Notre Dame Press**

Photography:

Satellite image usage courtesy of Environmental Research Institute of Michigan (ERIM).

Dave Carmichael, 104, 124, 131; Conference of Latin American Geographers, V; Richard Davis, 130; Val L. Eichenlaub, 13, 67, 75, 139; James French, 80, 94; Margaret Gloster, 102, 126; Jay R. Harman, 4, 6; H. Kruglak, "Barometric Autograph of a Tornado," 130; John Lemker, 4, 11, 36, 117, 125, 148, 158; Michigan Department of Agriculture/Climatology Program, 9; Hans J. Stolle, 150; WMU Weather Workshop, 88, 89, 95.

---


Manufactured in the United States of America

**Library of Congress Cataloging-in-Publication Data**

Eichenlaub, Val L., 1933–
The Climatic Atlas of Michigan / Val L. Eichenlaub, Jay R. Harman, Fred V. Nurnberger ;
cartographic director, Hans J. Stolle.
p. cm.
Includes bibliographical references.
ISBN 0-268-00773-X
1. Michigan—Climate—Charts, diagrams, etc. I. Harman, Jay R. II. Nurnberger, Fred V.
III. Stolle, H. J. (Hans J.) IV. Title.
QC984.M4E33 1990
551.69774′022′2—dc20 89-40020

# DEDICATION

*The Climatic Atlas of Michigan* is dedicated to Lucia C. Harrison, lifelong resident of Michigan, teacher in Michigan Public Schools and colleges for forty-five years, weather and climate enthusiast, and author of books on earth-sun relations.

Miss Harrison was born in Saginaw, where she taught in the public schools from 1898–1901. From 1903–1908, after receiving an AB degree from the University of Michigan, she was sixth grade supervisor at Northern State Normal, (now Northern Michigan University) at Marquette. It was in the remoteness and ruggedness of the Upper Peninsula that she developed an interest in geography and subsequently began course work at the University of Chicago. She began teaching at Western State Normal (now Western Michigan University) in 1909 and joined the Geography Department in 1912. She continued to teach in that department until her retirement in 1947.

Lucia Harrison received her MS degree at the University of Chicago in 1919, where her program included three courses in Meteorology and Climatology. That year, she joined the American Meteorological Society. By 1932 she had completed all coursework at the University of Chicago for the Ph.D. in Geography.

One of Miss Harrison's major interests was climatology. Her books reflected this interest as two of them dealt with earth-sun relations, the primary basis of climatic distribution. *Daylight, Twilight, Darkness and Time* was published by Silver Burdett and Co. in 1935, and *Sun, Earth, Time and Man* by Rand McNally Co. in 1960. These two works are classics, and required reading for understanding the driving force of climate, the distribution of energy from the sun.

After Lucia Harrison's death in 1974, her bequest to Western Michigan University's Geography Department made available an Endowment Fund. Over the years, the Endowment Fund has provided financial support for research projects, visiting speakers, graduate student fellowships, field studies, workshops, research publications, and countless other endeavors related to the enhancement of Geography at Western Michigan University.

Grants from the Lucia Harrison Endowment Fund have been the primary source of financial support for the planning, research, and cartographic work which has preceded the publication of *The Climatic Atlas of Michigan*. The atlas is a fitting tribute to Miss Harrison's long and productive teaching and writing career, and to her unflagging interest in weather and climate.

# TABLE OF CONTENTS

# LIST OF MAPS AND GRAPHS

# FOREWORD

*The Climatic Atlas of Michigan* is an example of successful collaborative research efforts among geographers who wanted to create both a comprehensive scientific record of our region's climate and a useful information resource for constituencies outside the scholarly community.

Professional geographers will find the atlas an important publication documenting weather and climate in Michigan. Others engaged in agriculture, commerce, or tourism will draw from the book valuable knowledge that may guide them in critical economic decisions. *The Climatic Atlas of Michigan* thus represents the ideal case in which solid academic scholarship and practical applicability meet and prove their dual value to the people of our state.

Very few other state climatic atlases exist. I am very proud that researchers of Western Michigan University have the privilege of adding another important and applicable research achievement to a long string of Michigana.

Diether H. Haenicke
President
Western Michigan University

# PREFACE

Climatic information has become imperative for a wide range of economic ventures, as well as for coping with the everyday necessities of life. With the goal to provide better climatic information for the citizens of Michigan, planning and preparatory work to produce a climatic atlas of Michigan began in the early 1980s. The project was conceived on the premise that a climatic atlas would be of great value to the general public, as well as to specialists in agriculture, transportation, tourism and recreation, utilities, planning, environmental sciences, engineering, and forestry.

This project has been a cooperative venture, combining the resources of the Geography Departments of Western Michigan University and Michigan State University and the Michigan Department of Agriculture/Climatology Program.

During early planning of the atlas content, a mutual decision was made by the atlas staff to provide climatic information not only in the form of maps of means and averages, but also of probabilities of occurrence. Thus, for many data items, the expected range of occurrence as well as average conditions are available for the atlas user. For the user's convenience, a decision was also made to present the data in the more familiar English units (Fahrenheit and inches) rather than in metric units with which the user may be less familiar. Conversion graphs are available on the maps.

The primary source of data used in the compilation of atlas maps has been that gathered by the National Weather Service network of climatological observers. The data were archived and summarized by the Michigan Department of Agriculture/Climatology Program. Using a base period of 1951–1980, the MDA/Climatology Program produced computer-plotted maps, which, after editing by the atlas staff, provided the basis for construction of the majority of the maps contained in the atlas. A plastic overlay map of the station network used in the compilation of these maps is provided in an envelope attached to the inside back cover of the atlas. It may be placed directly over the large maps of the state contained in the atlas to give the user more precise locational information.

Several preliminary research projects were deemed necessary to provide climatic information where summarized data were not already in existence. The results of these projects have been used for Sunshine and Cloudiness, Tornado, Fruit Belt, Pressure, and Climate Change sections of the atlas. In addition, data from several preexisting studies have been utilized in map or graphic mode in Wind, Solar Radiation, Evaporation, Thunderstorms, and Comfort Climate sections.

All cartographic work was done by the Atlas Cartographic Staff, Cartographic Services, Department of Geography, Western Michigan University. For the first time, a shaded relief map of Michigan has been constructed. This is included in the atlas. As Michigan weather is intimately tied to relief and topography, the map should serve as a guide in interpreting the details of many of the map patterns portrayed by the atlas.

Producing a climatic atlas for Michigan has been a challenging task. The spatial patterns of the many climatic variables are intricately related to lake proximity and to topography. Large variations may occur in short distances resulting in patterns of considerable complexity and detail. But this is, after all, what makes the climate of Michigan so interesting and the job of producing a climatic atlas a rewarding one.

# ACKNOWLEDGMENTS

The authors wish to extend thanks to many people, agencies, and organizations who, over the years, have lent support and encouragement in various ways to the Michigan Climatic Atlas project. We wish to extend special acknowledgments to the following: The Lucia Harrison Endowment Fund, Western Michigan University, for funding the majority of the preparatory, research, and cartographic work and a portion of the final printing costs; the Kalamazoo Consortium for Higher Education, the Undesignated/Unrestricted Fund of Western Michigan University, the Preparation, Presentation, and Publication of Papers and Exhibition of Creative Works Fund of Western Michigan University, and the Michigan Meteorological Resource Program—all of these for providing partial funding of final printing costs. We wish to acknowledge our debt to the Department of Geography, Western Michigan University, for their patience during the years and for providing cartographic facilities, and to the office staff of the Geography Department for many hours spent typing the atlas text. We would like to thank the Department of Geography of Michigan State University and the MSU computer facilities, which were utilized for data analysis and plotting; the Michigan Department of Agriculture, for their support in partial funding of the cartographic work and printing costs; the staff of the Michigan Department of Agriculture/Climatology Program for their work in preparing and analyzing data, in particular Maxine Oshel, Office Supervisor and jill-of-all-trades, Susan Perry, Computer Programmer, and the student part-time employees; Thomas Hodler who assisted in planning the atlas during its early stages; Bob Janiskee, for allowing part of his research on comfort climates to be used in the atlas; our many photo contributors; President Diether H. Haenicke of Western Michigan University, for his enthusiastic support of the project and for providing funds for the final printing.

We particularly want to thank the Western Michigan University's Cartographic Laboratory Staff, Christopher Forth, Douglas Kyle, Mary Dillworth, and Tom Haney, for the past three years' dedication to the atlas cartographic work. Our appreciation for their skills and talents and for their diligent work through the heat and humidity of three Michigan summers can never be adequately conveyed.

# INTRODUCTION

## The Setting

### *Geographical Features*

Michigan, located in the heart of the Great Lakes region, is composed of two large peninsulas. Many smaller peninsulas jut from these into the world's largest bodies of fresh water, which give most of Michigan a semi-marine type climate in spite of its midcontinent location.

The Upper Peninsula, located primarily between 45° and 47°N latitude, is long and narrow, and averages 75 miles in width. It lies between 84° and 90°W longitude, and extends from northern Wisconsin eastward more than 300 miles into northern Lake Huron. Lake Superior forms the boundary of the entire north shore, while Lake Michigan forms the boundary to the southeast. Isle Royale, separated from the mainland, is located in Lake Superior about 50 miles northwest of the tip of the Keweenaw Peninsula. This island, about 10 miles wide and 25 miles long, is a popular national park during the summer months.

The Lower Peninsula, shaped like a mitten, comprises about 70% of Michigan's total land area. It extends northward nearly 300 miles from the Indiana-Ohio border, which is around 42°N latitude, to the tip of the lower peninsula (the Straits of Mackinac), which is about 46°N latitude.

The Mackinac Bridge, spanning the Straits of Mackinac, joins the two large peninsulas where Lake Michigan meets Lake Huron.

Lake Michigan extends the entire length of the Lower Peninsula on the west, while Lakes Huron, St. Clair, and Erie and the connecting St. Clair and Detroit rivers form the eastern boundary. The total coastline of the state exceeds 3100 miles.

In addition, Michigan has over 11,000 smaller lakes scattered throughout 81 of the 83 countries. These lakes have a combined total surface area of over 1000 square miles. More than 36,000 miles of streams wind their way across the state.

Image for June 11, 1980 from data acquired by the Coastal Zone Color Scanner on board Nimbus-7 Satellite. Areas in red are vegetation.

### Topography

The terrain in the eastern half of the Upper Peninsula varies from level to gently rolling hills with elevations generally between 600 and 1000 feet above sea level. In the western Upper Peninsula elevations rise to between 1400 and 1600 feet. The state's highest point, approximately 1980 feet in Baraga County, overlooks Lake Superior. Rugged hills, geologically a series of hogbacks, extending northeastward from Ontonagon County through the center of the Keweenaw Peninsula, figure in the larger amounts of precipitation received in this area.

Lower Peninsula land-features range from quite level terrain in the Saginaw Lowland to gently rolling hills in the southwest and southeast, with elevations generally between 800 and 1000 feet. A series of sand dunes along Lake Michigan's shoreline rise nearly 400 feet above the mean lake level. These dunes are the result of prevailing westerly winds blowing across the lake. Higher terrain of glacial origin occurs in the northern part of the Lower Peninsula, reaching a maximum elevation of 1725 feet in Osceola County near Cadillac.

### Water Supply and Agriculture

Michigan is particularly fortunate in its supply of both surface and ground water. Surface water supplies are constantly replenished by an annual average precipitation of about 31 inches. Because of moderately high humidity and cool temperatures, evaporation rates are relatively low. Heavy industrial demands are made upon the water supply, but few industries have had to go any great distance to find adequate supplies. Aside from the availability of the lake water, industry can normally meet its water requirement needs at depths of 25 to 400 feet. There is, of course, an abundance of water to meet the needs of cities and individuals.

### Agriculture

Because of its climate, soils, and marketing conditions and large variation in agricultural practices, more than fifty different agricultural commodities are grown in Michigan. Ranked by dollars generated to farmers, the top fifteen are: milk, field corn, forest products, soybeans, beef, hay, pork, fruit, wheat, vegetables, dry edible beans, nursery items, potatoes, and sugarbeets.

The "fruit belt" is located in the southwestern and western borders of the Lower Peninsula along the shores of Lake Michigan where, due to prevailing westerly winds, the tempering influence of the lake water is strongest. Here, due to the special climatic features, commercial fruit production is concentrated.

### Recreation

Because of the influence of the Great Lakes on the climate, the variety of topographic features, and thousands of miles of lake shore, Michigan attracts a robust, year-round tourist business. Cooler temperatures and abundant natural beauty provide the opportunity for an excellent summertime vacation. The topography and the heavy snowfall are ideal for downhill skiing and tobogganing. Ice fishing, snowmobiling, and cross-country skiing are also popular activities. Michigan's climate provides the most favorable conditions in the Midwest for winter sport enthusiasts.

## Climatic Controls

Michigan's latitude is the major climatic control, determining the amounts and seasonal contrasts of incoming solar radiation (insolation). This accounts for the marked seasonal change which is a salient feature of Michigan's climate. However, the dynamic and changing character of Michigan's weather on a day-to-day basis is a result of the upper air circulation which occurs over the midlatitude portion of North America.

### Upper Air Circulation

Winds above 20,000 feet generate and steer daily weather systems that account for episodes of precipitation and fluctuations of temperature encountered during all seasons in the state. Ripples in this upper air circulation create traveling regions of atmospheric ascent and descent that generate centers of low and high atmospheric pressure ("cyclones" and "anticyclones," respectively) especially near the jet stream, the region of strongest high level winds. Dry, sunny weather is associated with descending air (an anticyclone), whereas wet weather is often associated with ascending air (a cyclone). These upper air "ripples" or waves normally cross the Midwest every two to four days and give Michigan's weather a certain periodicity that we come to expect—dry weather is followed by an interval of rain, which is then followed by another interval of dry weather.

The path followed by the upper atmospheric circulation and its associated cyclones and anticyclones may go unchanged for days or weeks, resulting in the weather crossing Michigan arriving from one characteristic source area. In this case, the weather may follow a monotonous pattern. When the circulation aloft has a strong northerly component and arrives in the Midwest from the Arctic, our weather averages much colder than normal and is dominated by air masses from that region. In winter, such a pattern causes bitter low temperatures but little precipitation except near the shores of the Great Lakes where lake effect snowfall may be heavy (see Snowfall Section). This same northerly upper level flow in summer would produce abnormally cool weather associated with polar air masses from Canada. On the other hand, southwesterly upper level flow brings storm systems into Michigan from the central and southern Great Plains that may pull with them large amounts of moist tropical air from the Gulf of Mexico. Such storms produce widespread outbreaks of snow or rain in winter and showers and thunderstorms at other seasons and

**FIGURE 1**
Typical jet stream situations in summer (above) and winter (below) (by H. J. Stolle).

account for much of the total precipitation across the state. At still other times, a large blocking upper level anticyclone may steer weather disturbances away from the Midwest, resulting in extended summer heat and drought or mild, stagnant winter weather. Fortunately, such blocked patterns are not frequent.

The exact location of the jet stream within the overall circulation pattern determines the path taken by cyclones and anticyclones as they cross the Great Lakes region. The more northerly position of the jet during summer generally confines most migratory cyclones to Canada and means that the Upper Peninsula, lying closer to the mean storm track, will have a more changeable summer climate than will the southern Lower Peninsula (figure 1). As the jet stream locates increasingly southward during autumn, all areas of the state see an increase in changeability and storminess, and by late winter sometimes the jet is located so far south of Michigan that episodes of storminess with widespread snow or rain are infrequent.

The strength of the upper air circulation and its associated jet stream determine the speed with which weather systems traverse the Midwest. The circulation is normally strongest in late winter and weakest in late summer. Weather features cross the region rapidly in the winter and sharp, sometimes dramatic changes may occur quickly as contrasting air masses are swept across the state by rapidly moving cyclones. In summer, however, the circulation is slower and features take more time to traverse the region, resulting in longer periods between weather changes.

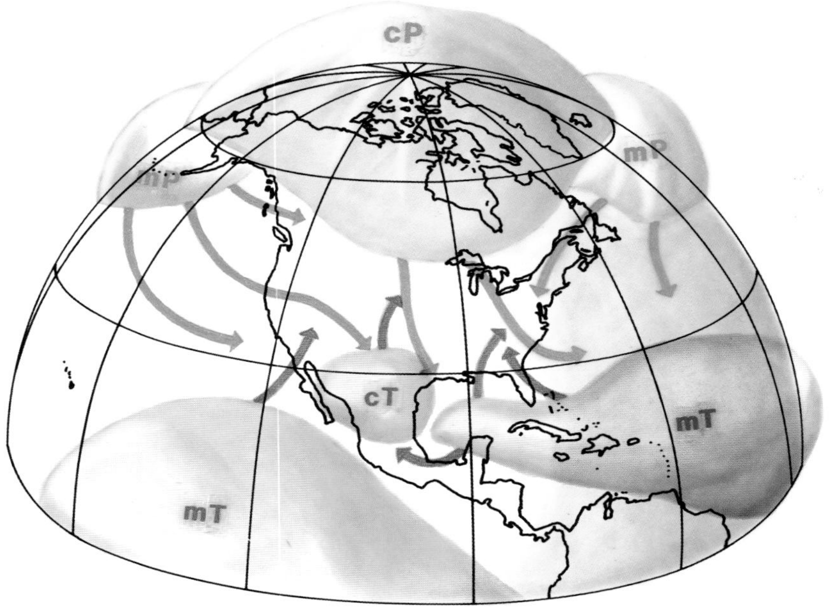


**FIGURE 2**
Typical air masses affecting North America and the Great Lakes area (by H. J. Stolle).

### Air Masses, Moisture and Humidity

Michigan weather is controlled much of the year by air masses originating on the North American continent. The Great Lakes augment the moisture content of these air masses slightly in winter (when the average water surface temperature is warmer than the air) but may actually remove some atmospheric moisture in summer when the water is significantly colder than the air (see following subsection on Lake Effects). Overall, the dominance of our weather by continental air masses and the limited, mainly seasonal impact by the Great Lakes mean that average relative humidity and dew point values across the state are only moderate, substantially lower than they might be at oceanic coastal location.

During winter, cold continental polar (cP) or milder Pacific (mP) air masses control our weather (figure 2). Both are characterized by low total water vapor content because they have either formed over a land mass or, in the case of Pacific air, traversed mountainous terrain and had a considerable amount of moisture already removed in the process. During the summer, tropical air masses from the Gulf of Mexico and Caribbean Sea become more frequent but still occur less often than the other air masses; tropical air has high moisture content and accounts for our spells of humid, muggy weather. Its greater frequency during this season helps to increase the average dew point and relative humidity values across the state. Also, the higher general temperatures of summertime polar air masses (compared to winter) allow them to hold more moisture as well, and they will have a higher dewpoint than in the winter.

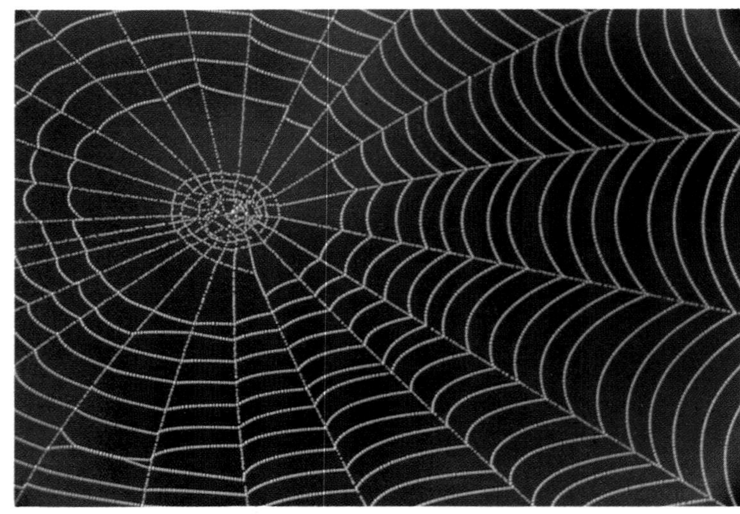

Dewy spider web.

## Lake Effects

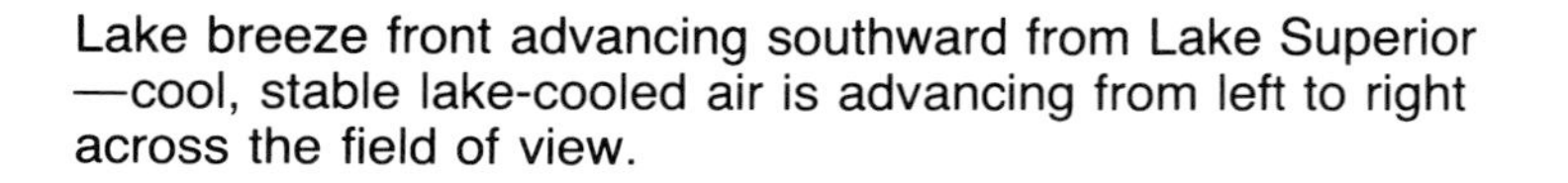

Lake breeze front advancing southward from Lake Superior —cool, stable lake-cooled air is advancing from left to right across the field of view.

Michigan is surrounded on three quadrants by the Great Lakes. These water bodies strongly modify the climate of Michigan at certain times and produce some weather twists that occur in few other areas of the world. Collectively, the array of semi-marine modifications imposed on Michigan's basically continental climate comprises the "lake effects" which are so much a part of Michigan weather.

Lake effects stem from three fundamental differences between the surrounding Great Lakes and the land surface of Michigan: (1) the lake temperatures "lag" behind the land temperatures with large differences occurring at certain times of the year; (2) the lakes increase the availability of moisture to be evaporated into the air during the cold season; (3) the surfaces of the lakes are smoother than the land.

### The Lag of Temperature between Land and Lake

The Great Lakes which surround Michigan warm more slowly than the land in the spring and cool more slowly in the fall (figure 1). Consequently there are two periods during the year where the mean lake-land temperatures may differ significantly. During these times strong thermal modifications may be imposed by the lakes on the surrounding shores.

In the spring, the land begins to warm rapidly as a result of increased daily totals of solar radiation. The lake surface temperatures warm much more slowly so that the mean lake temperatures may be colder than the mean land temperatures during much of the period March–August. The lakes require more energy than the land to warm, utilize some solar energy for evaporation, transmit solar energy to greater depths as compared to opaque land materials, and, most importantly, are constantly in motion mixing warmed surface water with colder water from the depths. Density differences within the lakes also contribute to slow spring warming. Water is most dense at 4° Celsius (39.2° F). Hence, when spring warming occurs at the surface, these waters become more dense, sink, and are replaced by colder waters from the depths. The entire column of water must be warmed to maximum density before rapid warming can occur.

In the fall when surface waters cool, they become

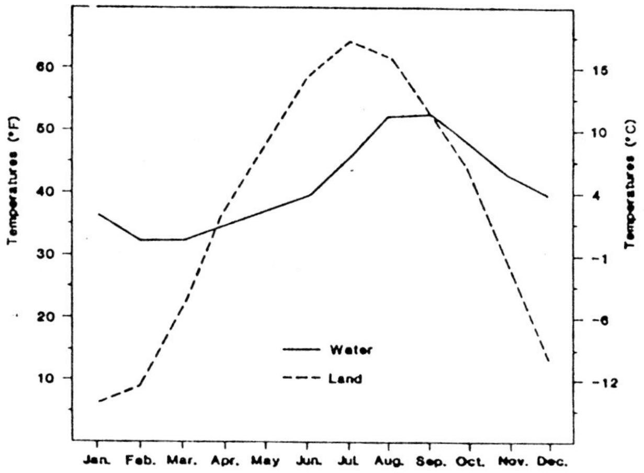


**FIGURE 3**

Mean water vs land perimeter temperatures, Lake Superior, from V. L. Eichenlaub, "The Effect of Lake Erie on Climate," in *The Great Lake Erie* (Ohio State University Research Foundation, 1987), p. 32. Data from D. W. Phillips "Environmental Climatology of Lake Superior" *Journal of Great Lakes Research* 4 (3–4, December 1978): 291.

more dense and sink, being replaced by warmer subsurface water. Surface cooling is retarded and the immense heat storage of the lake keeps temperatures above those of the surrounding land during much of the time from September–March.

During both spring and fall seasons the lake-land temperature differences can be very large. When very warm air masses move over Michigan in the spring, the temperature contrast between lake and land can be as much as 40° F and the air is chilled. In the late fall and winter, on certain occasions, temperature contrasts can be even larger with lake temperatures warmer than air temperatures. During these periods the air is warmed.

## *Effects of the Lakes on Moisture Availability*

When air passing over the lakes is cold relative to the surface temperature of the lakes, as during fall and winter, the evaporation rates may be large and the lakes may be locally important sources of moisture to the air over Michigan. When the air is warmer than the lakes, little evaporation occurs and the air may actually lose moisture to the lakes (condensation on the lake surface). Evaporation rates are usually low during the spring and early summer.

## *Effects on Winds Resulting from Smoother Surfaces of the Lakes*

Winds are altered in direction and in velocity as they flow from rougher land to smoother lake surface or vice-versa because of differences in friction induced by the contrasting roughness of lake and land surfaces and the subsequent effect on speeding or slowing the winds. Velocity changes in turn alter the coriolis force which is a partial determinant of wind direction. When winds flow from land to lake, velocity increases and a bending to the right occurs as the coriolis effect increases. When the winds flow from the smoother lake to the rougher land the opposite usually occurs.

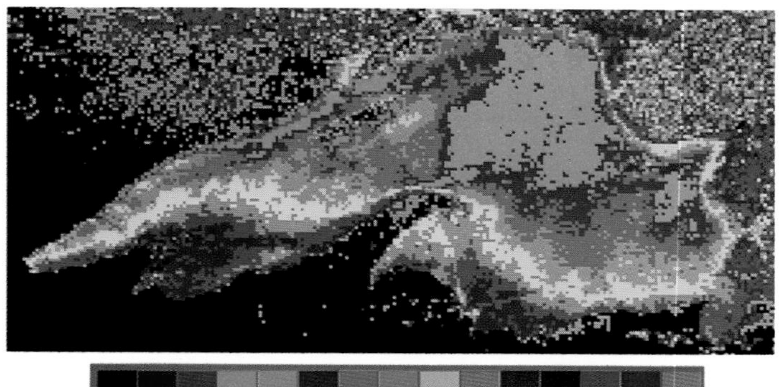


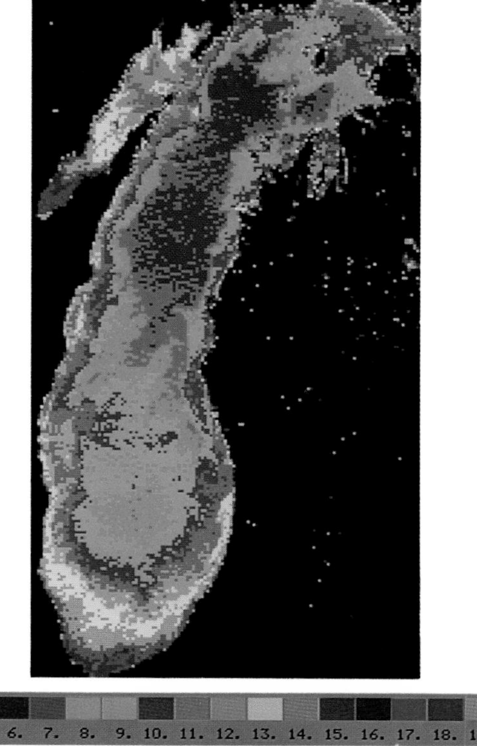


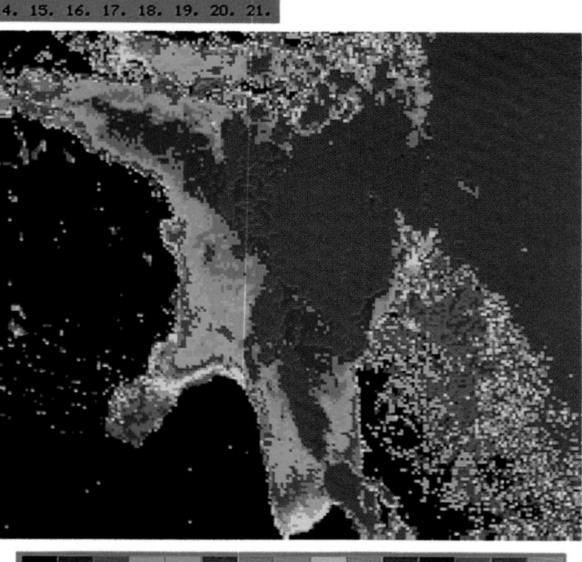


Surface water temperatures (C°) for Lake Superior, Lake Huron, and Lake Michigan. Images produced from data obtained by Coastal Zone Color Scanner, on board Nimbus-7 Satellite, June 11, 1980.

## Spring and Summer Lake Effects

During much of the time from about March through August lake temperatures are lower than land temperatures and evaporation rates are low. The vertical motions that occur as a result of heating of the earth's surface are suppressed over and near the lakes with the result that some types of clouds are suppressed and the lakeshore environment may enjoy more sunshine than inland areas.

The thermal effects of cooler lake waters restrict the mean daily range of temperature along the lake shore, while reducing the mean daily temperatures (see Temperature Section). Most marked is the reduction in the daily maximum along the immediate lake shore. While daily minimum temperatures are less strongly affected, they are generally increased. The last frost of the spring comes slightly earlier along the shores of the lakes although the increased length of the growing season is chiefly due to delayed frost and resulting extension of the growing season during the fall months. In some portions of the Lake Superior Basin the difference in the frost free season between interior and lake shore areas may be as much as two or three months.

Lake breezes are of common occurrence in shore areas during spring and summer. These breezes blow from the lake to the land as a result of small pressure differences that develop daily as the land warms but the air over the lakes remains relatively cold. Lake breezes may spring up during the late morning, reach a peak in the afternoon, and then die out toward late afternoon. They may cause sharp reversals of wind direction, cause a cooling of the normal midday heat, and increase the relative humidities.

During the spring and summer, fogs may develop over the lakes and along shorelines as a result of the marked cooling of moist air as it passes over cold lake waters. The fogs may be brought some distance inland by lake breezes but usually dissipate when warmer land areas are reached. Fogs along the Lake Superior shoreline may occur more often during the March-August period due to persistently cold surface waters throughout the summer.

Lake effect snowsquall over Lake Michigan.

## Fall and Winter Lake Effects

From September through February the lakes are generally warmer than the land and the effects on climate are quite different than during the spring and summer. The lakes warm the air and increase the vertical motion which may cause cloud development. As a result of this warming and the high frequency of cyclonic storms that cross Michigan in winter, Michigan's winters are among the cloudiest in the United States (see Sunshine and Cloudiness Section).

The lakes cause warmer temperatures in lee shore areas with the most marked effect on the daily minimum temperature (see Temperature Section). The daily temperature range is restricted and the first frost of fall occurs considerably later along the shores than in interior areas. In general the modifications on temperature imposed by the lakes during the cool season are stronger than those during the warm season and extend further away from the lakes.

In fall and winter, the lakes may provide important amounts of moisture to the air. Combined with vertical motion enhanced by heat from the lakes, amounts of snowfall on the lee shores of Lake Michigan and Lake Superior may be much heavier than on windward shores (see Precipitation Section). "Lake effect" snow (resulting from the overpassing of warm lake waters by cold air) is a familiar part of Michigan's winter weather complex, and probably the most prominent of the lake effects (figure 4).

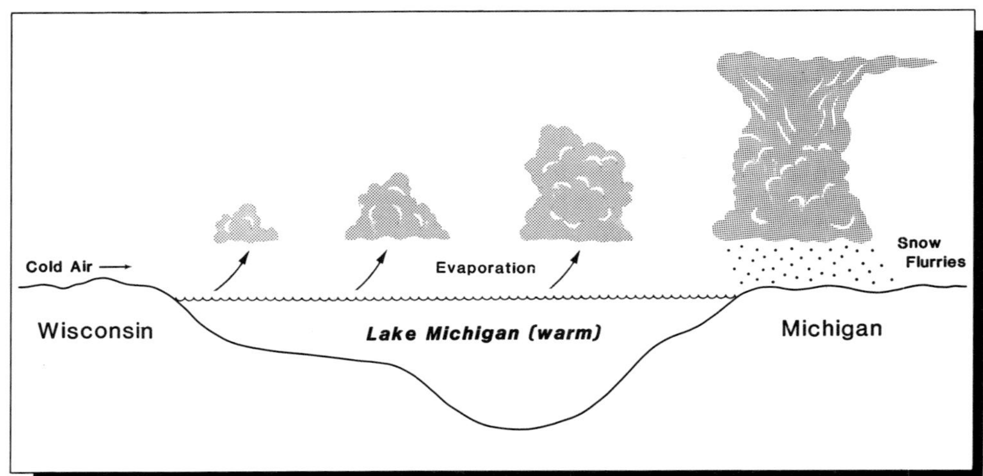


**FIGURE 4**
Development of lake effect snow as cold air crosses warm waters of Lake Michigan.

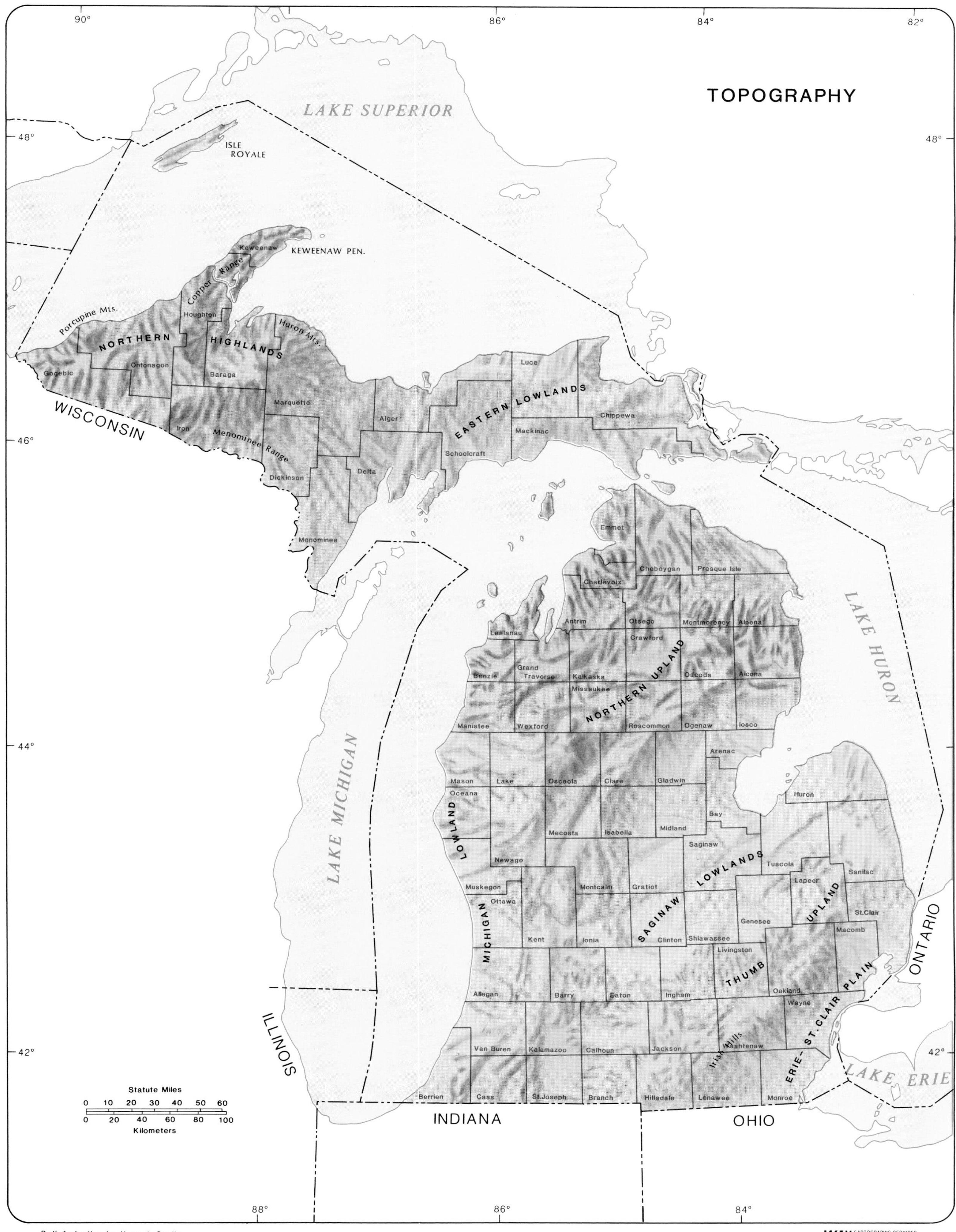


Relief shading by Hans J. Stolle

WMU CARTOGRAPHIC SERVICES DEPARTMENT OF GEOGRAPHY

# DATA

## Sources and Use

The data used for this atlas have been made available in response to the July 1, 1891, Act of Congress, as amended, which has assigned to the National Weather Service (NWS) the task of taking and maintaining such meteorological observations as may be necessary to establish and record the climate conditions in the United States. To accomplish this assigned task, the National Weather Service, now a part of the U.S. Department of Commerce/National Oceanic and Atmospheric Administration, provides weather instruments to volunteer cooperative observers who record weather data for their respective location so their own and future generations may benefit from the accumulation and use of these data. The authors wish to acknowledge and thank the hundreds of cooperative observers for the many years of dedication to provide the data. In addition, thanks to the staff of the National Climatic Data Center (NCDC) for their work in providing high quality data to the end users after it has been received from the cooperative observers.

The historical NWS climatological network consists of approximately one station for each 625 square mile area. From this network, the state climate program for Michigan, within the Michigan Department of Agriculture,

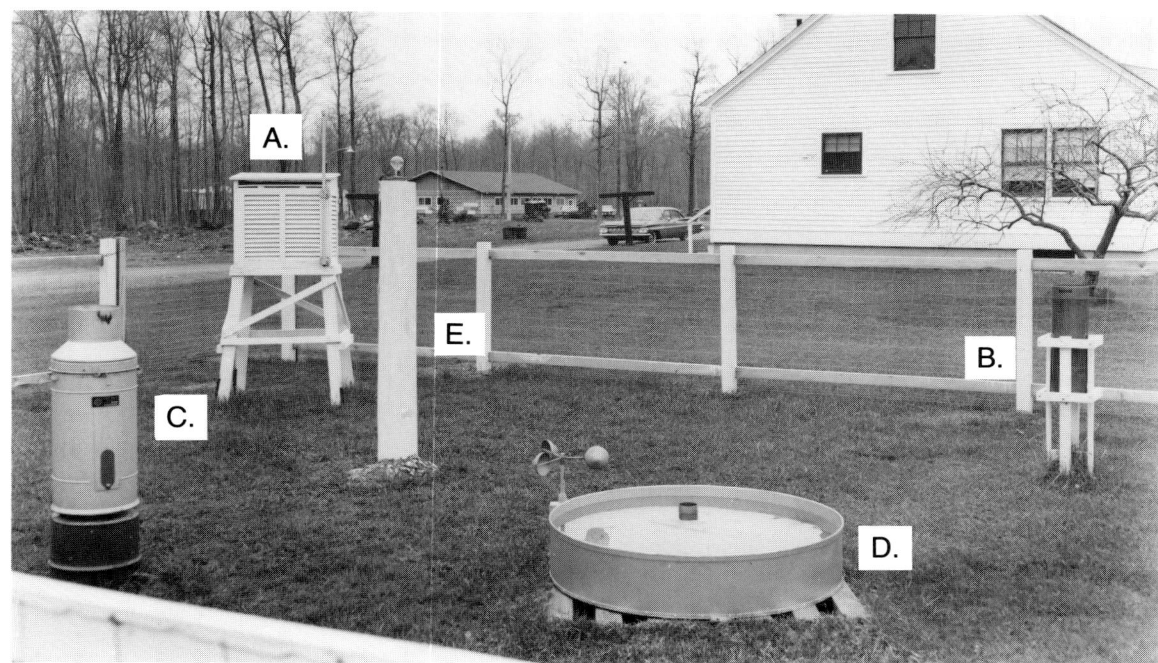


**An example of a nearly complete climate station:**

A. Standard National Weather Service thermometer shelter containing the maximum and minimum thermometers.

B. Standard National Weather Service 8 inch rain gauge.

C. 10 inch universal weighing rain gauge.

D. Standard National Weather Service "Class A" evaporation pan which includes (1) stilling well with hook gauge, (2) totalizing 3-cup anemometer, (3) maximum and minimum water temperature thermometers, (4) support platform to break contact with the ground.

E. Pyranometer to measure total incoming thermal radiation.

Missing: soil temperature thermometers and anemometer at 2 or 6 meters.
Most cooperative observers have A and B only.

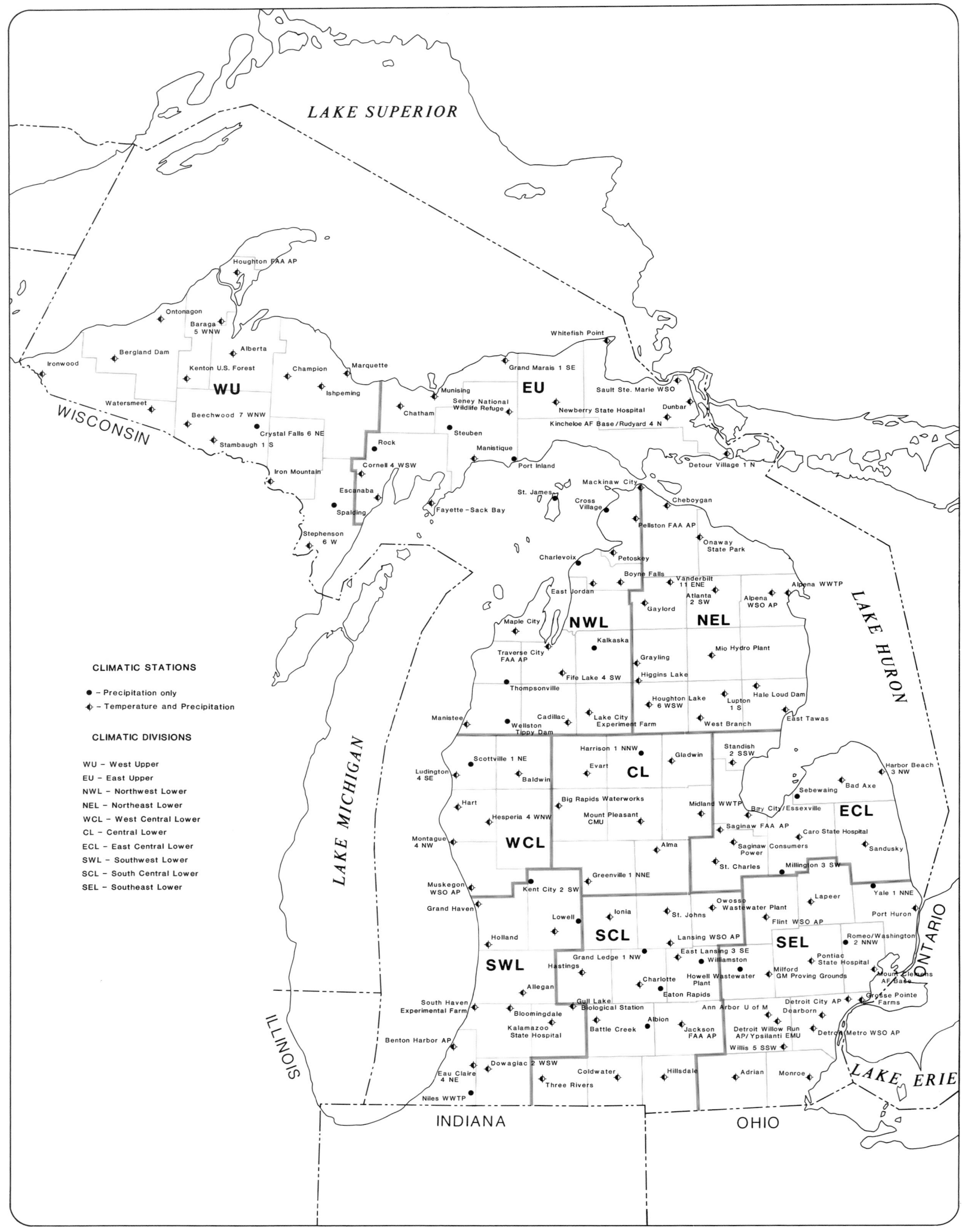
LAKE SUPERIOR
LAKE MICHIGAN
LAKE HURON
LAKE ERIE
WISCONSIN
ILLINOIS
INDIANA
OHIO
ONTARIO
WU
EU
NWL
NEL
WCL
CL
ECL
SWL
SCL
SEL
CLIMATIC STATIONS
● - Precipitation only
- Temperature and Precipitation
CLIMATIC DIVISIONS
WU - West Upper
EU - East Upper
NWL - Northwest Lower
NEL - Northeast Lower
WCL - West Central Lower
CL - Central Lower
ECL - East Central Lower
SWL - Southwest Lower
SCL - South Central Lower
SEL - Southeast Lower
Houghton FAA AP
Ontonagon
Baraga 5 WNW
Alberta
Bergland Dam
Ironwood
Kenton U.S. Forest
Champion
Marquette
Ishpeming
Watersmeet
Beechwood 7 WNW
Crystal Falls 6 NE
Stambaugh 1 S
Iron Mountain
Whitefish Point
Grand Marais 1 SE
Munising
Chatham
Seney National Wildlife Refuge
Sault Ste. Marie WSO
Newberry State Hospital
Dunbar
Kincheloe AF Base / Rudyard 4 N
Rock
Steuben
Manistique
Port Inland
Cornell 4 WSW
Detour Village 1 N
Escanaba
Spalding
Stephenson 6 W
Fayette - Sack Bay
St. James
Mackinaw City
Cross Village
Cheboygan
Pellston FAA AP
Onaway State Park
Charlevoix
Petoskey
Boyne Falls
Vanderbilt 11 ENE
Atlanta 2 SW
Alpena WWTP
Alpena WSO AP
East Jordan
Gaylord
Maple City
Kalkaska
Traverse City FAA AP
Grayling
Mio Hydro Plant
Fife Lake 4 SW
Higgins Lake
Thompsonville
Houghton Lake 6 WSW
Lupton 1 S
Hale Loud Dam
Manistee
Cadillac
Wellston Tippy Dam
Lake City Experiment Farm
West Branch
East Tawas
Harrison 1 NNW
Gladwin
Standish 2 SSW
Scottville 1 NE
Evart
Ludington 4 SE
Baldwin
Harbor Beach 3 NW
Bad Axe
Sebewaing
Big Rapids Waterworks
Hart
Midland WWTP
Bay City/Essexville
Hesperia 4 WNW
Mount Pleasant CMU
Saginaw FAA AP
Caro State Hospital
Montague 4 NW
Saginaw Consumers Power
Sandusky
Alma
St. Charles
Millington 3 SW
Greenville 1 NNE
Muskegon WSO AP
Kent City 2 SW
Yale 1 NNE
Grand Haven
Owosso Wastewater Plant
Lapeer
Ionia
St. Johns
Port Huron
Lowell
Flint WSO AP
Holland
Lansing WSO AP
Romeo/Washington 2 NNW
East Lansing 3 SE
Grand Ledge 1 NW
Williamston
Pontiac State Hospital
Hastings
Charlotte
Howell Wastewater Plant
Milford GM Proving Grounds
Mount Clemens AF Base
Allegan
Eaton Rapids
South Haven Experimental Farm
Gull Lake Biological Station
Detroit City AP
Grosse Pointe Farms
Ann Arbor U of M
Dearborn
Bloomingdale
Albion
Kalamazoo State Hospital
Battle Creek
Jackson FAA AP
Detroit Willow Run AP/Ypsilanti EMU
Detroit Metro WSO AP
Benton Harbor AP
Willis 5 SSW
Dowagiac 2 WSW
Coldwater
Hillsdale
Adrian
Monroe
Eau Claire 4 NE
Three Rivers
Niles WWTP

has obtained the daily computerized data for all stations for the period 1951–1980. The choice of this period of record is not arbitrary, but is in conformance with the NCDC's adherence to the definition of a "climatological normal" period established by the World Meteorological Organization. The period of record for normals is the most recently completed tri-decadal period. To aid in filling in some gaps in the station network of 1951–1980 stations, all stations with 20 years or more of record during the 1951–1980 period have also been included.

The daily computerized data have been reviewed, best estimates supplied for missing data (if three months or less were missing for a given station in a given year), and obvious errors replaced. Various internal consistency checks were made by computer programs, and data failing the checks were flagged for inspection and resolution of the problem by the validators. The resulting "cleaned" data were summarized for various times, statistical analyses were performed, and the results used to produce the following maps. Exceptions to this procedure and/or data set and time period are noted in the relevant sections. The resulting data set was for 111 stations that recorded both temperature and precipitation and 25 additional stations that recorded only precipitation. The stations used are shown on the accompanying map and on the plastic overlay provided in the back of the atlas.

Upon inspection of the station location map, one can readily see there are many areas of the state that have significant gaps in the data. This is due to two reasons. First, the NWS network does not have uniform spacing because of population distribution and the willingness of volunteers to cooperate. Second, the length of record may not be long enough to qualify a station for inclusion in the analysis. In some cases, critical stations were not usable due to incompatible moves in the middle of the record. A prime example is the Grand Rapids NWS Office. It was moved from the Old Kent County Airport, which was close to the downtown area, to the new Kent County International Airport, a more rural setting, on November 24, 1963. Since this was right in the middle of the 30-year period, neither location qualified with 20 or more years of record. The authors have attempted to properly interpolate the areas where data were sparse. Only time and additional data will prove whether the efforts were correct or not.

The daily data used have several characteristics that should be noted by the user:

First, the thermometers are a maximum and minimum type that are read to a precision of 1°F and reset once per day. They are housed in a naturally ventilated instrument shelter to shield them from the influences of solar radiation and precipitation. The thermometers are positioned at a nominal height of 5 feet above the ground surface. No corrections have been made for depth of snow cover as it relates to the actual height of the thermometer above the "surface."

Second, it is well known that due to the nature of the thermometers and the various times of the day that the different observers take their observations and reset the thermometers to the current temperature, there is a systematic bias in some of the data. For example, if an observer takes the reading at 4 p.m., the thermometers will be set at the current air temperature. Since the maximum thermometer will not change unless the air temperature the next day is higher, the reading the next day will have a carry over temperature that is a false maximum if the true maximum temperature was cooler. Likewise, the reverse will be true for an early a.m. observation. If the next day's minimum is not as cold as the reset temperature that day, the recorded minimum will be a carry over false minimum. The major problem in correcting for these biases is the large number of different times of the day that various observers take their readings and the limited number of stations that have hourly temperatures for comparison. No adjustment for this "time of observation" bias was used for this analysis.

Winter sunrise, Milham Park, Kalamazoo.

Third, liquid precipitation is measured in a NWS standard 8 inch diameter rain gauge to a precision of .01 inch. Snowfall is measured to a precision of .1 inch at a location that, at the observer's discretion, is representative for the area. The effects of the wind in a wide open area or around buildings, trees, and other obstructions make this very difficult if not impossible at times. A prescribed procedure is then followed to obtain a sample of the new snowfall for that day to be melted to determine the liquid content. It is the liquid content that is included in the yearly total of precipitation. In addition, the time of day that an observer makes the observation affects the amount of snowfall measured. For example, a 5 p.m. observation on a warm day or several hours after the snow has stopped falling will record less snowfall due to melting and settling than an observation taken in the morning or closer to the time the snowing stopped. A midnight observer, although recording the data for the true 24-hour day, will also be low for early morning or midday snowfall but have a more accurate reading of late afternoon and evening events. NWS and Federal Aviation Administration (FAA) stations that take hourly observations will record more snowfall by summing the hourly readings than will the once-a-day observers. This is especially true of the light events and overnight trace amounts that daytime only observers will miss. No adjustments have been made for these various biases. The above cautions are only intended to alert the user to many of the uncertainties that are inherent in data of this type. The data used are the best available.

Fourth, the NWS network density provides a good representation of the macro to meso, i.e., large to medium, scale climate areas in Michigan. Because climatic variables change more rapidly near the Great Lakes and where there are significant changes in altitude, care must be exercised when applying the data from a single point to a larger area.

Fifth, because climate changes most rapidly near the ground and due to local topography, soils, and other factors, i.e., the micro-climate, one must be very careful in extrapolating the data to near surface conditions or to a specific location that was not observed. The information provided herein should be used for general planning purposes only. More specific details about a given area should be obtained by supplemental monitoring as the cost/benefits warrant.

Questions regarding additional information concerning the climatic features of Michigan or specific station information may be obtained upon request to the Michigan Department of Agriculture/Climatology Program.

# Probabilities

A brief word about the difference between the average or mean, and 50th percentile values. The reader may be most familiar with the "normal" probability distribution. This has the common "bell-shaped" symmetrical frequency curve. In the "normal" distribution, the mean value is, by definition, also the 50th percentile value. Many natural data can adequately be described by the "normal" distribution, e.g., average monthly or annual temperatures, dates of freezes, growing season lengths, and others. Some natural data, however, are not adequately described by this distribution. These data, for example, precipitation, snowfall, degree days, number of days of a certain event, and others, do not have a symmetrical distribution. They have instead a skewed distribution whereby a longer tail of data exists on one side of the most frequently occurring value, i.e., the mode. If this tail extends to the right of the graph it is said to have positive skew and the opposite if the longer tail is to the left.

Precipitation and degree days have a positive skew since a few large values can exist but no values less than zero are possible. A major portion of the data is grouped around the smaller values. Thus the average will be increased by the large values and be greater than the 50th percentile. The 50th percentile means half of the values are larger and half are smaller which is, also, the definition of the median of the data. Therefore, the 50th percentile value for a skewed distribution is more meaningful since it implies that half the time there will be more and half the time less which is not the case for the average.

## Probability Distributions Used in Atlas

Throughout this atlas, if the "average" or "mean" value is given plus probabilities, then a *normal* distribution was used. If the 50th percentile value is given, then a *Gamma* distribution was used.*

The presence of a zero value on the 50% probability map *does not* imply that the event has never occurred in that region. Instead, the interpretation should be that in half the years or more in the 30-year period, it did not occur. Conversely, if a value does appear on the 50% probability map, it does not imply that every year averages that amount, but it does say that half the years are that amount or less. The reader may obtain more specific details by contacting the State Climatologist at the MDA/Climatology Program office.

---

*With many of the data types for which the Gamma distribution was used, values of less than zero are not valid. Furthermore, inclusion of the zero values directly in the calculations creates problems in calculations if the percentage of zero was very large, for example, spring and fall transition months. Therefore, to account for the zero values and have meaningful results, a compound model was used. The implementation was as follows. The number of years with zero were converted to a fractional value of the record at the station. For example, if the station had all 30 years of record but 18 were zeros, then the fractional value was .6. Thus, during the 30-year period there is a 60% chance of zeroes for the element type and time at that station. The non-zero data were analyzed separately and appended to this probability of a zero value. Thus, in the above example a 75% probability would be comprised of the 60% due to zero and the 15% probability based on the non-zero values. In this model, the calculation for any probability less than or equal to the probability of zero results in a value of zero. Hence, the 50th percentile value in the above example would be 0.

# TEMPERATURE

## Introduction

Temperature is one of the most important features of climate. It controls the range of outdoor activities that people engage in, determines human comfort and the energy expenditure to create comfortable indoor climates, sets limits to vegetation growth and agricultural production, and affects the occurrence of precipitation in either the liquid or solid forms. In addition, temperature interacts with other weather elements in countless ways.

The maps of average temperature in Michigan show some surprisingly large differences within short distances. These patterns arise because Michigan temperatures respond to a rather complex interaction of factors, rather than to a single control.

The most important factor is latitude, which determines the daily angle of the sun and length of day. That, in turn, determines the balance (net radiation) between incoming solar radiation and outgoing terrestrial radiation. When incoming solar radiation exceeds outgoing terrestrial radiation (positive net radiation) surfaces warm. When outgoing terrestrial radiation exceeds incoming solar radiation (negative net radiation) surfaces cool.

The seasonal contrasts in net radiation within the state cause large differences between average winter and summer temperatures. Seasonal temperature contrasts are also accentuated by Michigan's location near the interior of North America. Temperatures cool rapidly in the fall and warm rapidly in the spring, as do large land masses. There are also significant north-south temperature gradients resulting from latitudinal gradients of net radiation.

While latitude and continentality explain a large part of the temperature pattern, other factors also affect Michigan's temperatures. The Great Lakes cool and warm more slowly than the surrounding land (see Lake Effects subsection). Their effects on temperature have been previously described. Secondary but important local factors such as snowcover, topography, and urbanization also affect Michigan temperatures. When snow covers the ground, temperatures are usually colder. Snow is a good insulator, and little heat transfer occurs from the ground into the atmosphere. Snow also reflects much of the incoming solar radiation, lessening net radiation values and causing colder temperatures. Most important, however, is the effect of snowcover on nocturnal temperatures. Snow is a very effective radiator of infrared (terrestrial) radiation, and consequently a snowcovered surface cools rapidly at night and experiences much lower minima than a surface free of snow. Thus, the northern portions of the state experience lower temperatures not only due to smaller positive and larger negative net radiation values, but also to a longer and more durable snowcover.

Standard instrument shelter to house maximum and minimum thermometers.

Maximum and minimum thermometers being read and reset.

Although topographic differences within the state are not great, the effect of local topography on temperatures may be very important at some locations. As air cools it becomes denser and accumulates in low areas. Thus, stations located in topographic depressions experience colder minimum temperatures. However, a method to quantify the local topographical influence by itself has not been developed. The effect of topography on maximum temperatures within the state is small.

Cities are usually warmer than their surrounding rural areas, particularly at night. The artificial materials of the city store more heat than the natural materials of the countryside. This heat can be released into the atmosphere causing the existence of a "heat island". This is a zone of higher temperatures developing under certain conditions, particularly on clear, windless nights. Where temperature sensors are located in an urbanized area, the existence of "heat islands" may have an effect on temperature averages, causing slightly higher values.

The maps in the Temperature Section of the atlas are for the most part based on computer maps provided by the MDA/Climatology Program. The data were observed and recorded by the National Weather Service Network of observers and archived by the MDA/Climatology Program.

### *Temperature Statistics*

The following definitions are provided to assist the user in the interpretation of the temperature maps:

*Daily Mean* = the daily maximum + daily minimum divided by 2

*Average Daily Mean* (*Monthly*) = the summation of the *Daily Means* divided by the number of days in the month

*Average Annual Daily Mean* = the summation of the *Average Daily Means* (*Monthly*) divided by 12

*Average Daily Maximum* (*Monthly*) = the summation of the daily maxima divided by the number of days of the month

*Average Annual Daily Maximum* = the summation of the *Average Daily Maxima* (*Monthly*) divided by 12

*Average Daily Minimum* (*Monthly*) = the summation of the daily minima divided by the number of days of the month

*Average Annual Daily Minimum* = the summation of the *Average Daily Minima* (*Monthly*) divided by 12

# The Annual March of Temperature

The annual march of temperature in Michigan is typical of a midlatitude continental location. Although the Great Lakes moderate temperatures somewhat, summers are warm, winters are cold, and the annual range of temperature is large.

Graphs for Marquette County Airport, Sault Ste. Marie, Alpena, Houghton Lake, Muskegon, Lansing, and Detroit Metropolitan Airport show average daily maximum, average daily mean, and average daily minimum temperatures for each month of the year over the period 1951–1980. The source of data was the publication *Local Climatological Data, Annual Summary*.

July is, at all six stations, the warmest month. The average daily maximum for July ranges from 83.1°F at Detroit to 75.1°F at Sault Ste. Marie while the average daily mean ranges from 71.9°F at Detroit to 63.5°F at Sault Ste. Marie.

January is the coldest month based on average daily maximum and mean temperatures. The average daily maximum ranges from 30.6°F at Detroit to 20.1°F at Marquette. The average daily mean varies from 23.4°F at Detroit to 12.1°F at Marquette. However, February is likely to be the month with the coldest minimum temperatures, particularly in the north. At this time, the extent of ice is at a maximum on the Great Lakes and clear nights are more common than in January. Average daily minimum temperatures at Sault Ste. Marie are 5.4°F in January and 5.3°F in February. At Houghton Lake the respective monthly values are 8.7°F and 8.0°F, while at Alpena they are 8.5°F and 8.0°F. The coldest absolute minimum temperatures in Michigan are usually recorded in February, although there are some exceptions.

The annual range of temperature (the difference between the average mean of the warmest month and the average mean of the coldest month) varies from 47°F at Muskegon to 52.5°F at Marquette. In general, the annual range becomes larger toward the north and west.

The daily range of temperatures is smallest in winter (only 11.4°F at Muskegon for December) and largest in summer (23.3°F at Lansing in July).

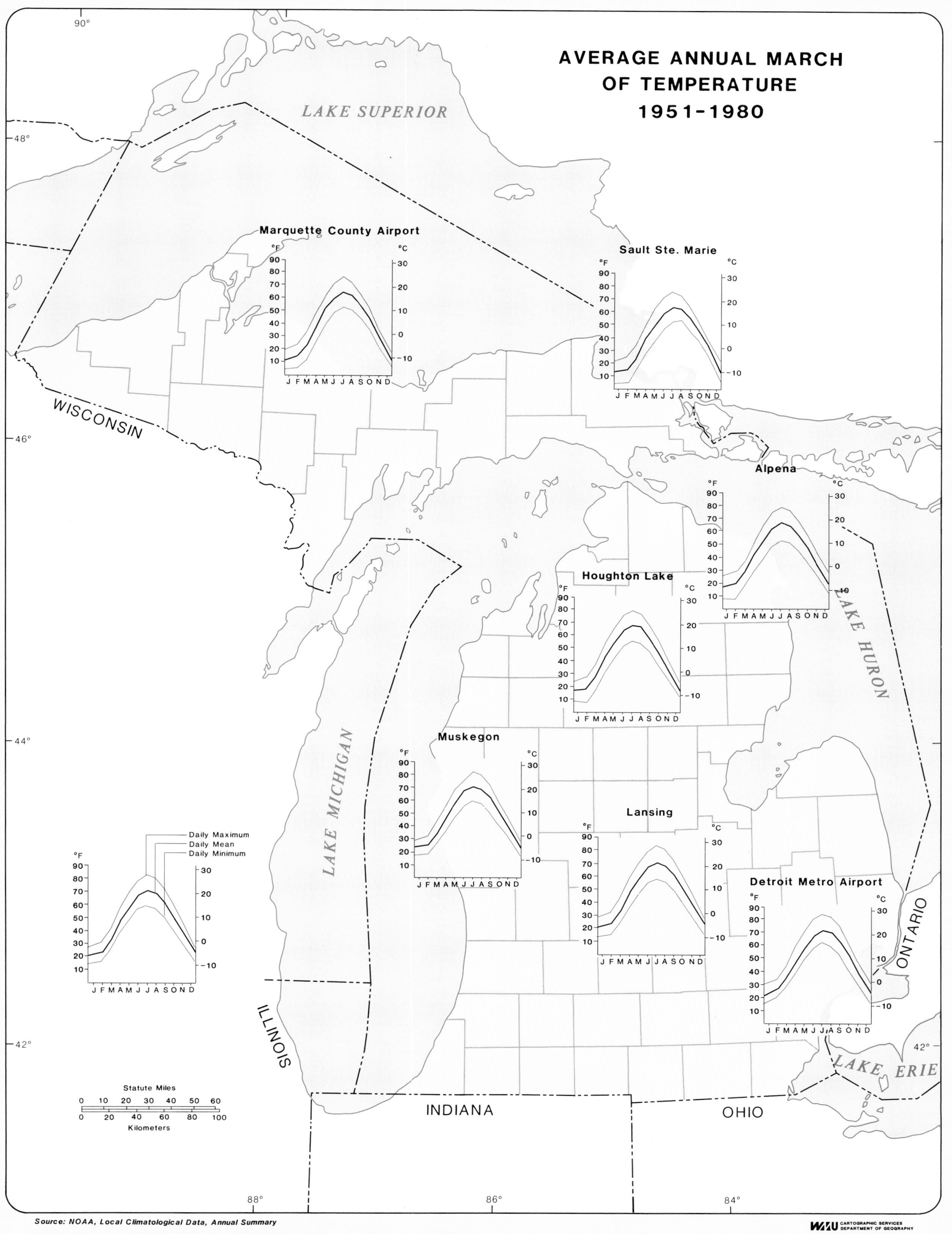


Source: NOAA, Local Climatological Data, Annual Summary

WMU CARTOGRAPHIC SERVICES DEPARTMENT OF GEOGRAPHY

# Average Daily Mean Temperatures

## *Average Annual Daily Mean Temperatures*

The map of average annual daily mean temperatures provides a rough approximation of comparative temperatures in Michigan and serves as an introduction to the temperature climatology of the state.

The map pattern reflects latitude, lake effects, and to a small degree, urbanization. The largest values (slightly less than 50°F) occur in the extreme southeastern portion of the state. The Detroit urbanized area, due to its southern location and to the heat island effect which elevates daily minimum temperatures, experiences the warmest average annual daily means.

The lowest values occur in the interior of the western portion of the Upper Peninsula (Champion 38.8°F, Bergland Dam 39.0°F) and near Whitefish Bay in the eastern portion of the Upper Peninsula (Sault Ste. Marie 39.7°F, Whitefish Point and Rudyard 39.9°F).

The effect of the Great Lakes on the annual average daily mean temperatures is to increase them. This is primarily the result of the fairly significant increase of daily minimum temperatures caused by the lakes during the winter.

## *Average Daily Mean Temperatures (Monthly)*

The maps of monthly average daily mean temperatures in Michigan show patterns strongly controlled by latitude and the effects of the lakes. In general, a south to north gradient exists during all months, but the map patterns change seasonally. During the winter, the role of the lakes is to elevate average daily mean temperatures, primarily by raising nighttime minima. Thus the lowest daily means are located away from the lakes in interior portions of the western Upper Peninsula. Bergland Dam has the somewhat dubious distinction of having the lowest daily mean temperatures during the winter months, December through March. In January, the daily mean temperature at Bergland Dam is only 10°F, the lowest value for any month for any station.

By April, the Great Lakes begin to cool surrounding shores and lower average daily mean temperatures, primarily by lowering the daytime maxima. The cold pole in the state shifts from the interior of the western Upper Peninsula to the shore of Lake Superior. From April through August, Whitefish Point, surrounded on three quadrants by the cold waters of Lake Superior and Whitefish Bay, averages the lowest daily mean temperature in the state. In July, the average daily mean temperature at Whitefish Point is only 61.1°F.

The warmest average daily mean temperatures occur in the extreme southeast and southwest portions of the Lower Peninsula. In the southeast, the effects of urbanization combine with latitude to produce the warmest daily means during many months of the year.

In January, the highest average daily mean temperature in the state occurs at Grosse Pointe Farms, with 25.0°F. In July, the warmest station is Detroit City Airport with 73.7°F. This is also the warmest monthly average daily mean experienced for any station.

The largest contrasts in average daily mean temperatures occur during the height of winter and in early summer. In January a 15°F contrast exists between the coldest and warmest stations, Bergland Dam and Grosse Pointe Farms. In June, Whitefish Point is 14.5° colder than Detroit City Airport. In the winter, mean daily temperatures are depressed by low nighttime minima occurring in snowcovered interior locations of the Upper Peninsula. In June, the lake plays a strong role in depressing daytime maxima along the shores of Lake Superior thus lowering the daily mean. On the other hand, in September and October the contrasts are small. In October, the difference between the coldest stations (Bergland Dam and Houghton, 45.0°F) and the warmest station (Grosse Pointe Farms, 53.9°F) is only 8.9°F.

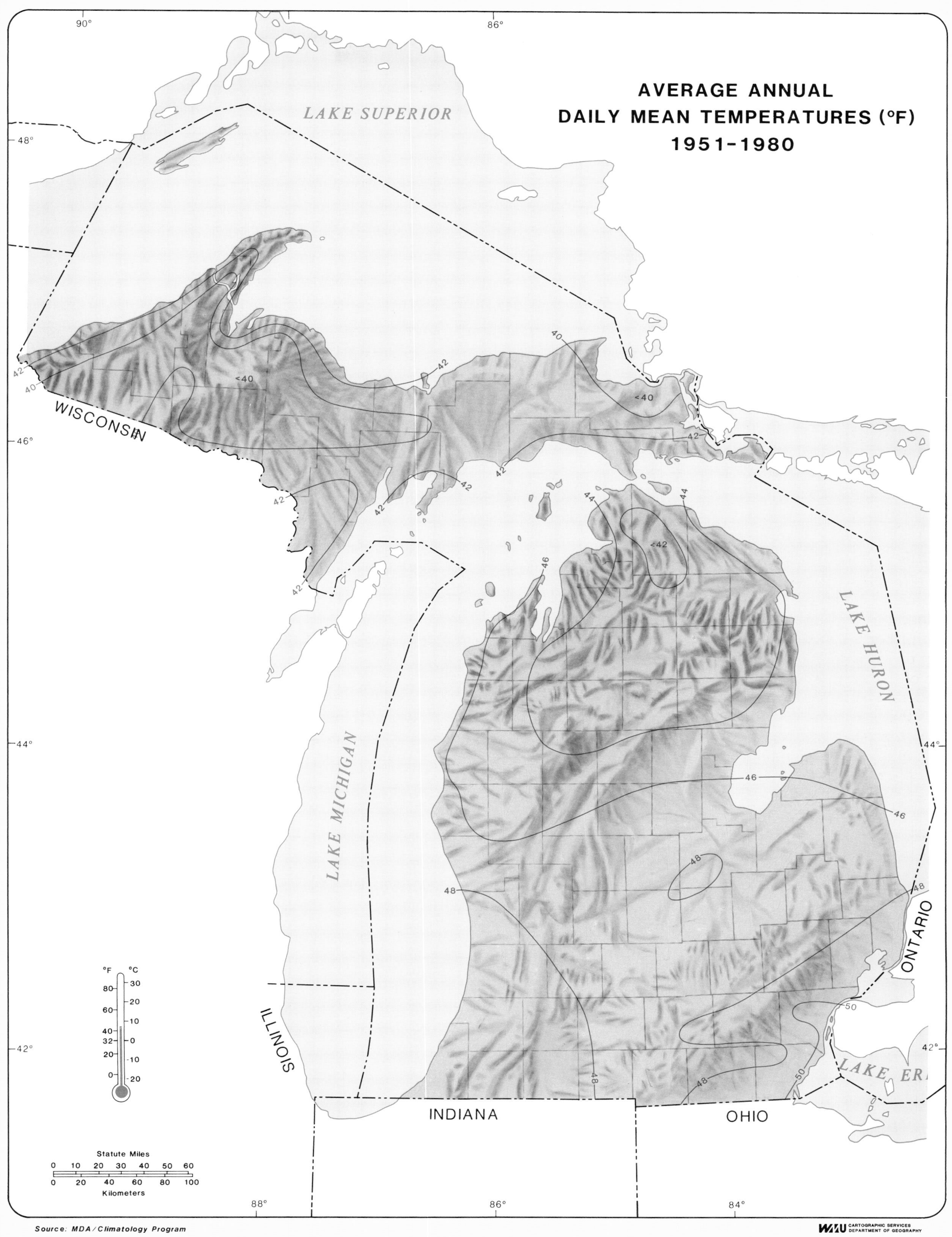


Source: MDA/Climatology Program

WMU CARTOGRAPHIC SERVICES DEPARTMENT OF GEOGRAPHY

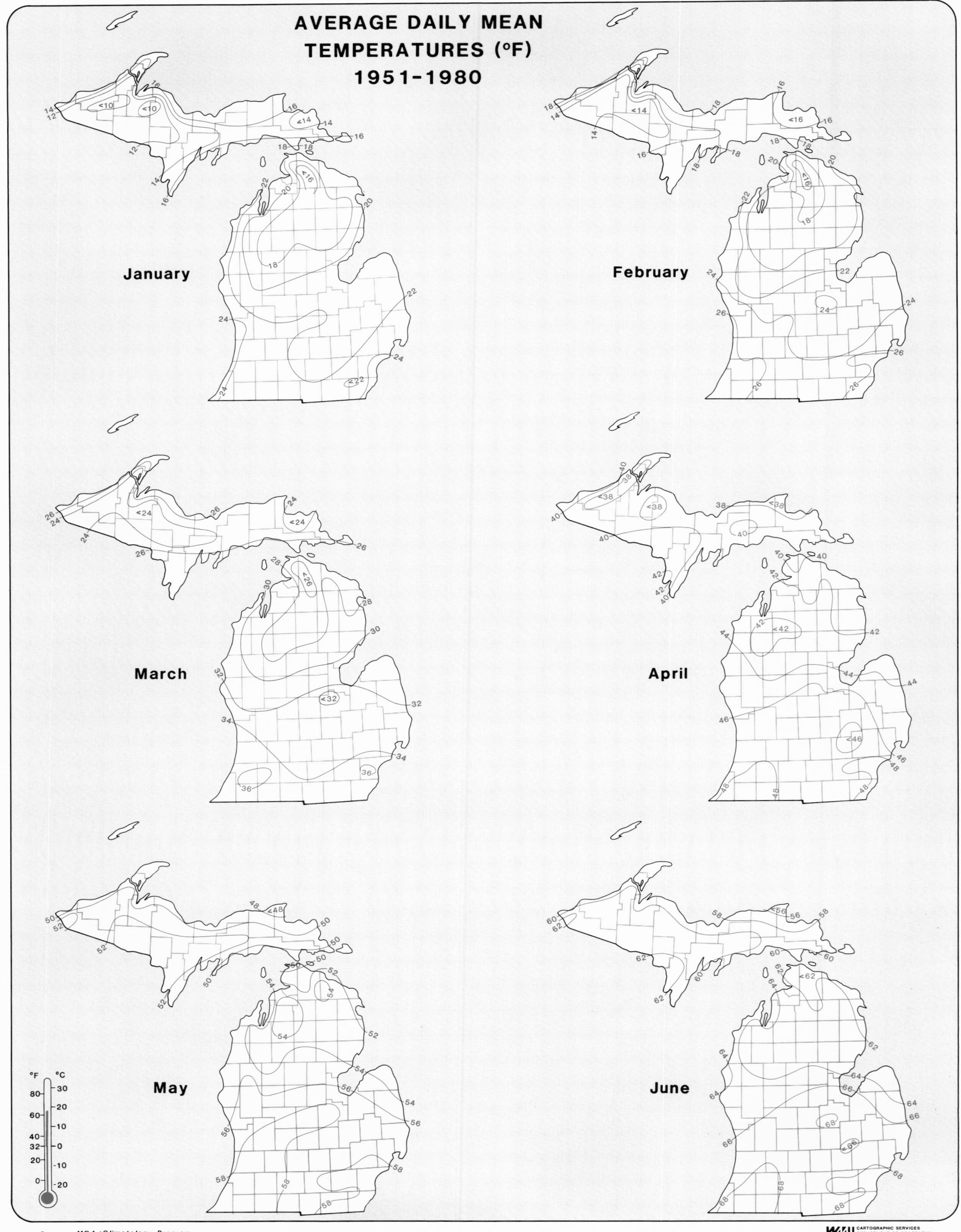


Source: MDA/Climatology Program

WMU CARTOGRAPHIC SERVICES DEPARTMENT OF GEOGRAPHY

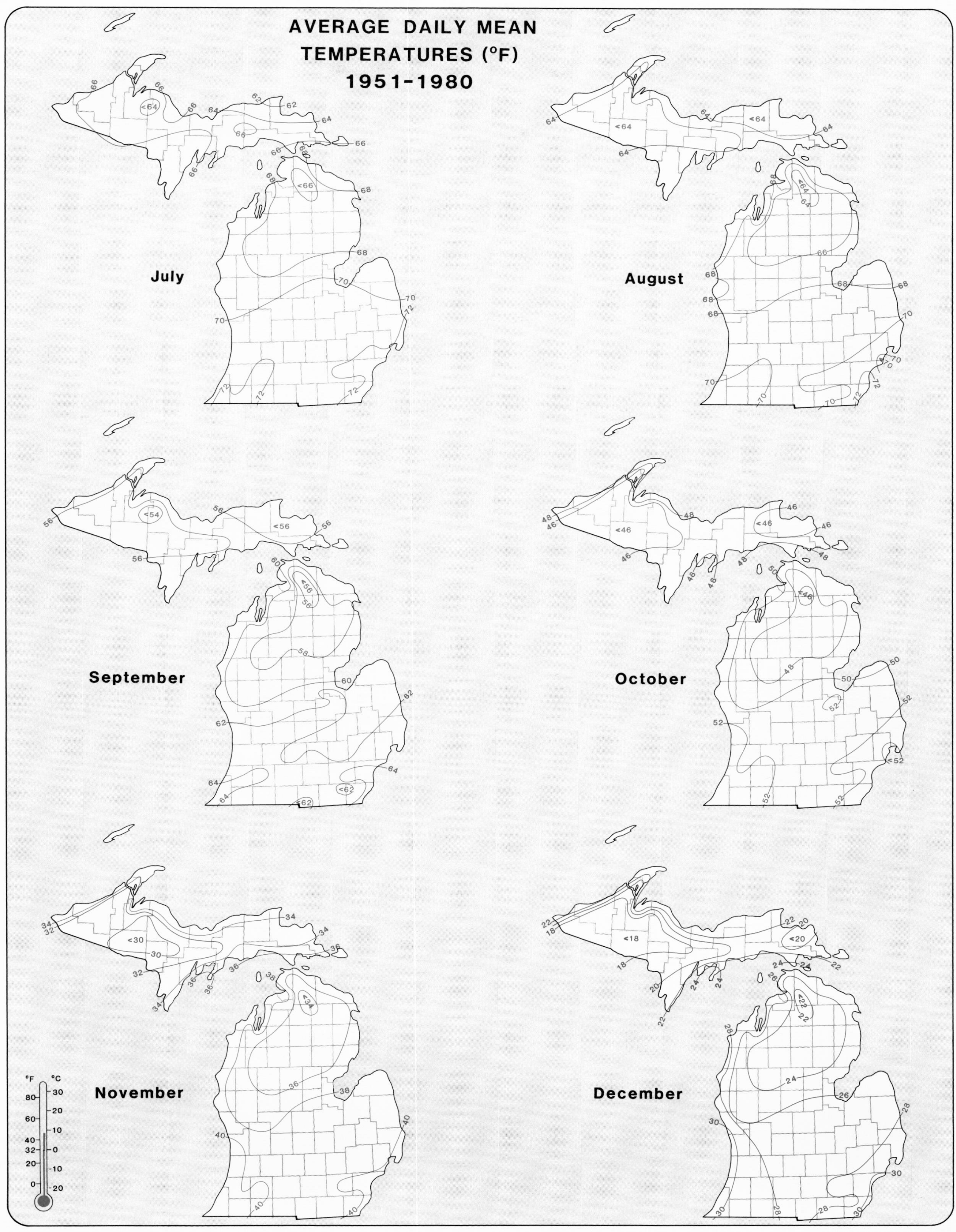


Source: MDA/Climatology Program

WMU CARTOGRAPHIC SERVICES DEPARTMENT OF GEOGRAPHY

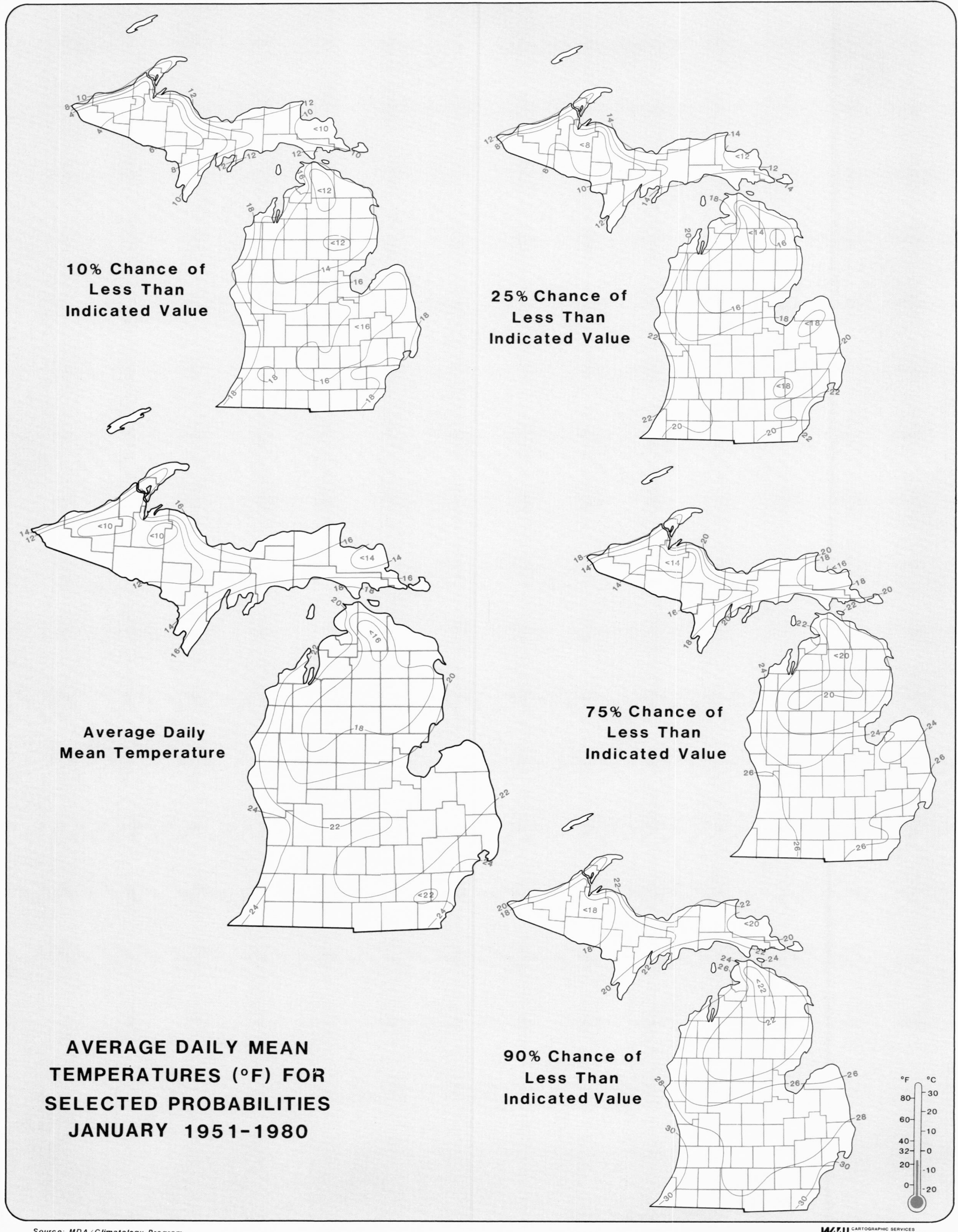


AVERAGE DAILY MEAN TEMPERATURES (°F) FOR SELECTED PROBABILITIES JANUARY 1951-1980

Source: MDA/Climatology Program

WMU CARTOGRAPHIC SERVICES DEPARTMENT OF GEOGRAPHY

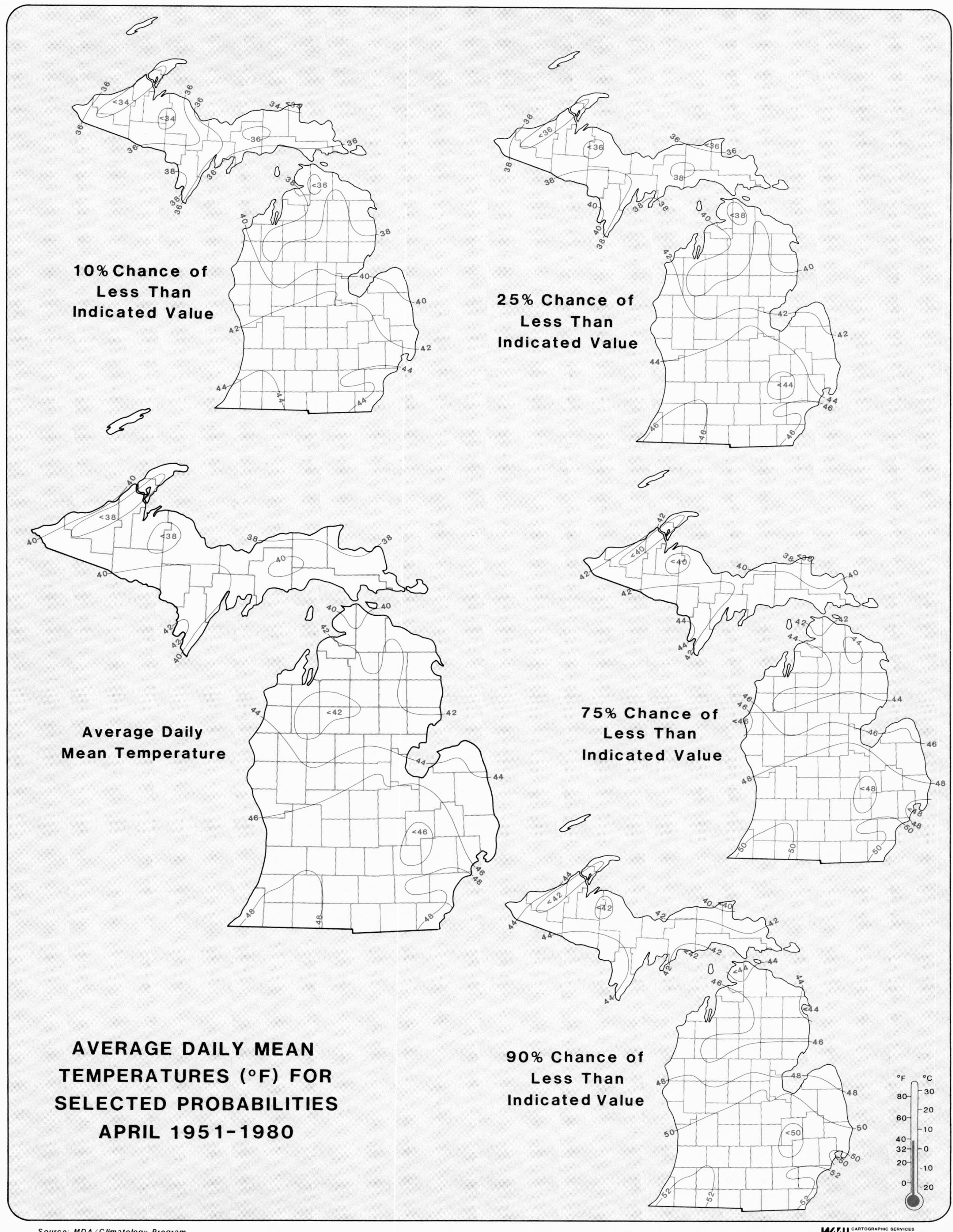


Source: MDA/Climatology Program

WMU CARTOGRAPHIC SERVICES DEPARTMENT OF GEOGRAPHY

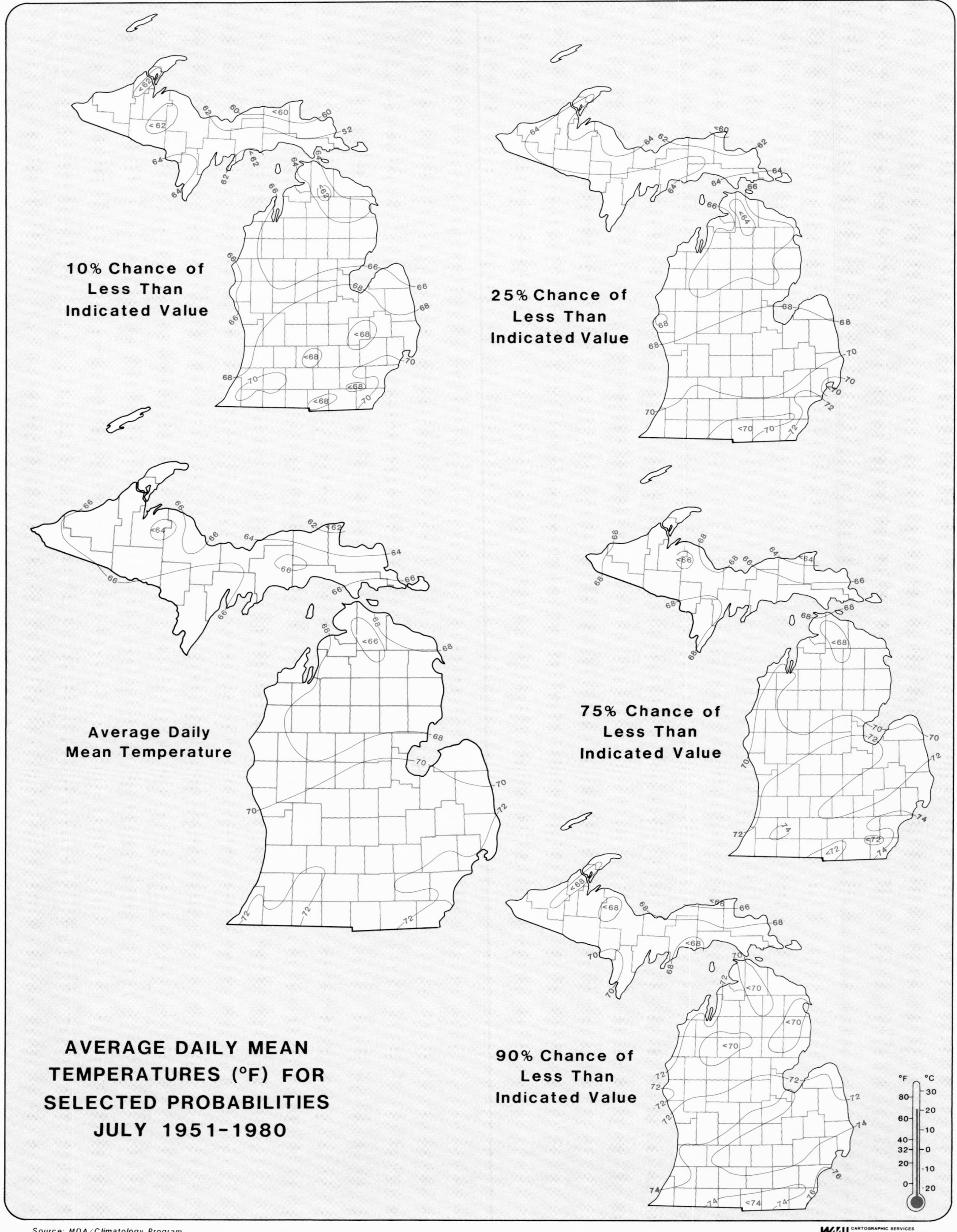


*Source: MDA/Climatology Program*

WMU CARTOGRAPHIC SERVICES DEPARTMENT OF GEOGRAPHY

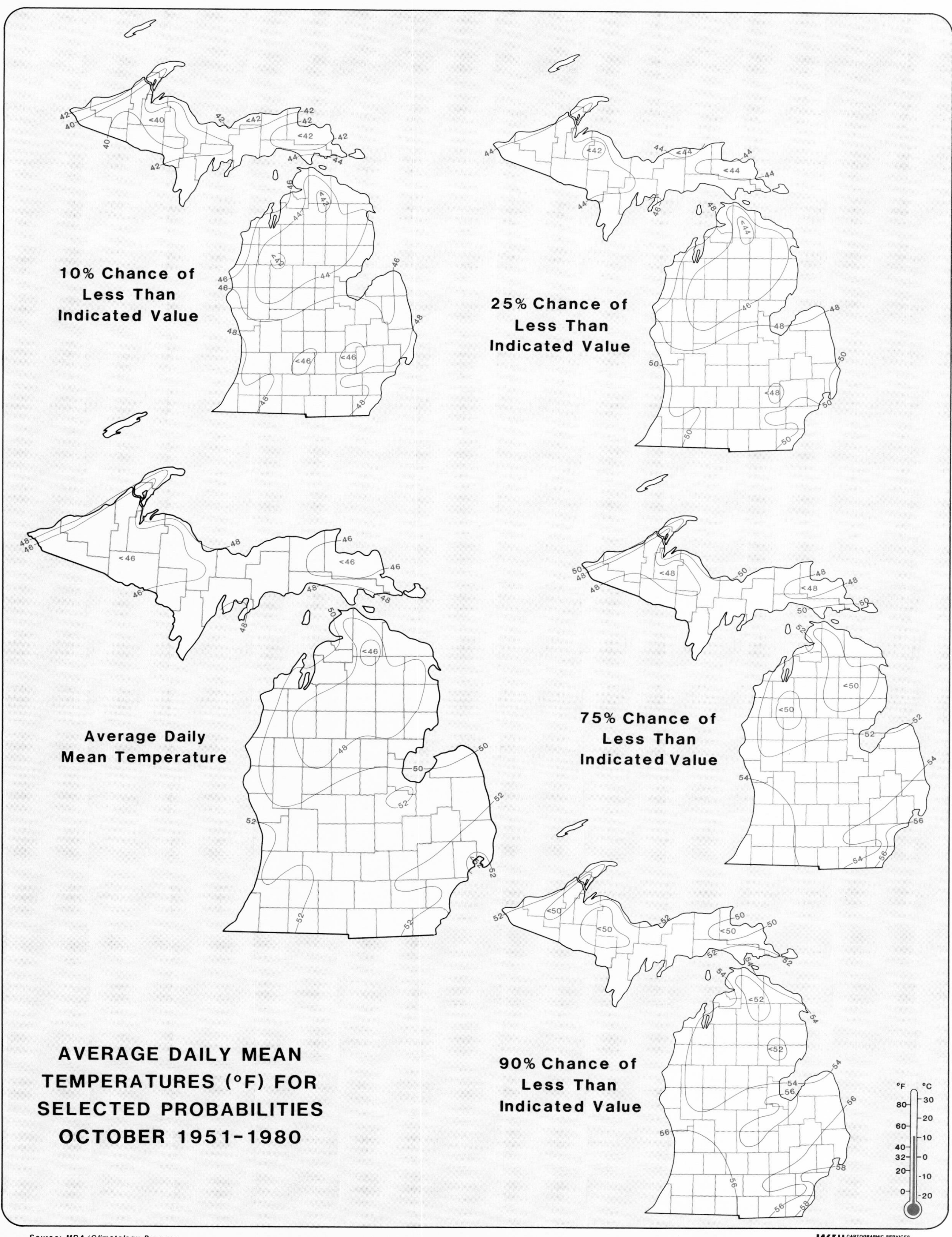


*Source: MDA/Climatology Program*

WMU CARTOGRAPHIC SERVICES DEPARTMENT OF GEOGRAPHY

# Average Daily Maximum Temperatures

## *Average Annual Daily Maximum Temperatures*

Averaged for the entire year, the daily maximum temperatures (normally the daytime highs) show the expected latitudinal decrease from south to north. Lake effects, which raise daily maxima during the cool season and lower them during the warm season, have a canceling effect and thus are not prominent controls for the annual map.

The warmest average annual daily maxima occur at Dearborn (59.8°F) and at Dowagiac (59.6°F). The values decrease by over 10 degrees northward to the Lake Superior shoreline. The lowest value occurs at Houghton with 47.6°F.

## *Average Daily Maximum Temperatures (Monthly)*

Michigan's latitude and continentality, resulting in strong seasonal changes experienced by the state, lead also to large monthly differences in average maximum temperatures. The lowest value occurs in January at Houghton and Ironwood with 19.8°F. The highest values occur in July at Kalamazoo and Dearborn with 84.9°F and 84.8°F respectively.

In January, average daily maximum temperatures are below freezing over the entire state with the exception of the extreme southwestern and southeastern corners of the Lower Peninsula. The lowest average daily maxima, 20°F or lower, occur on the Keweenaw Peninsula and in the Ironwood area of the extreme western Upper Peninsula.

In contrast to minimum temperatures, maximum temperatures warm substantially in February. Stations in the southern half of the Lower Peninsula normally experience daily highs above freezing. The cold pole of average maximum temperature shifts to the eastern Upper Peninsula as the western interior of the Upper Peninsula begins to warm somewhat due to its greater continentality. Houghton (22.1°F) has the coldest average maximum and Detroit City Airport and Dowagiac have the warmest (34.9°F).

By March, above freezing average daily maxima occur at all stations except Houghton. The coldest daytime highs occur on the Keweenaw Peninsula (Houghton 31.0°F) and the warmest in the southwestern part of the state at Kalamazoo (45.3°F) and Dowagiac (45.0°F). April brings average maxima of 60°F or higher in extreme southwestern counties of the Lower Peninsula, but daily highs average only about 44°F on the Keweenaw Peninsula. During April the lake effect begins to suppress daytime highs along the shores of the Great Lakes.

In May, daily maxima average above 70°F in many southern counties, and exceed 60°F over much of the Upper Peninsula except for the Keweenaw Peninsula and Whitefish Point, both surrounded by the cooling waters of Lake Superior. Kalamazoo is the warmest at 72.3°F and Whitefish Point the coolest at 55.7°F.

June average daytime highs rise above 80°F in some interior portions of the southern Lower Peninsula, and over 70°F elsewhere except for the immediate Lake Superior shore. Kalamazoo remains the warmest at 81.1°F and Whitefish Point remains coolest at 65.3°F. July, the hottest month, sees maxima above 80°F over the majority of the interior of the Lower Peninsula, and above 84°F over counties in the extreme southwest. Coolest daily maxima occur on the Keweenaw Peninsula and Whitefish Point, where they fail to average above 72°F.

In August, the 80°F isotherm retreats southward to the central portion of the Lower Peninsula and by September 80°F average daily maxima disappear from the state. Daily highs in the Upper Peninsula and northern interior of the Lower Peninsula do not average 70°F or higher. Kalamazoo and Detroit City Airport have the warmest maxima (76.1°F), while Houghton is coolest with 62.8°F. The effect of the Great Lakes on daytime highs diminishes in September as the lake surface temperature is near its maximum and land areas begin to cool.

By October, only the southern half of the Lower Peninsula normally experiences a daytime high of 60°F or above. The northern portion of the Lower Peninsula and all of the Upper Peninsula see daytime maximum temperatures averaging only in the 50s. In November, daytime highs average less than 50°F over the entire state and less than 40°F in portions of the Upper Peninsula. Dearborn has the highest average maximum, 49.7°F, with Houghton the lowest at 36.7°F.

In December, daily maxima average below freezing over the Upper Peninsula and northern interior Lower Peninsula. The highest average monthly maximum occurs at Grosse Pointe Farms with 36.9°F, and the lowest at Houghton with 25.3°F.

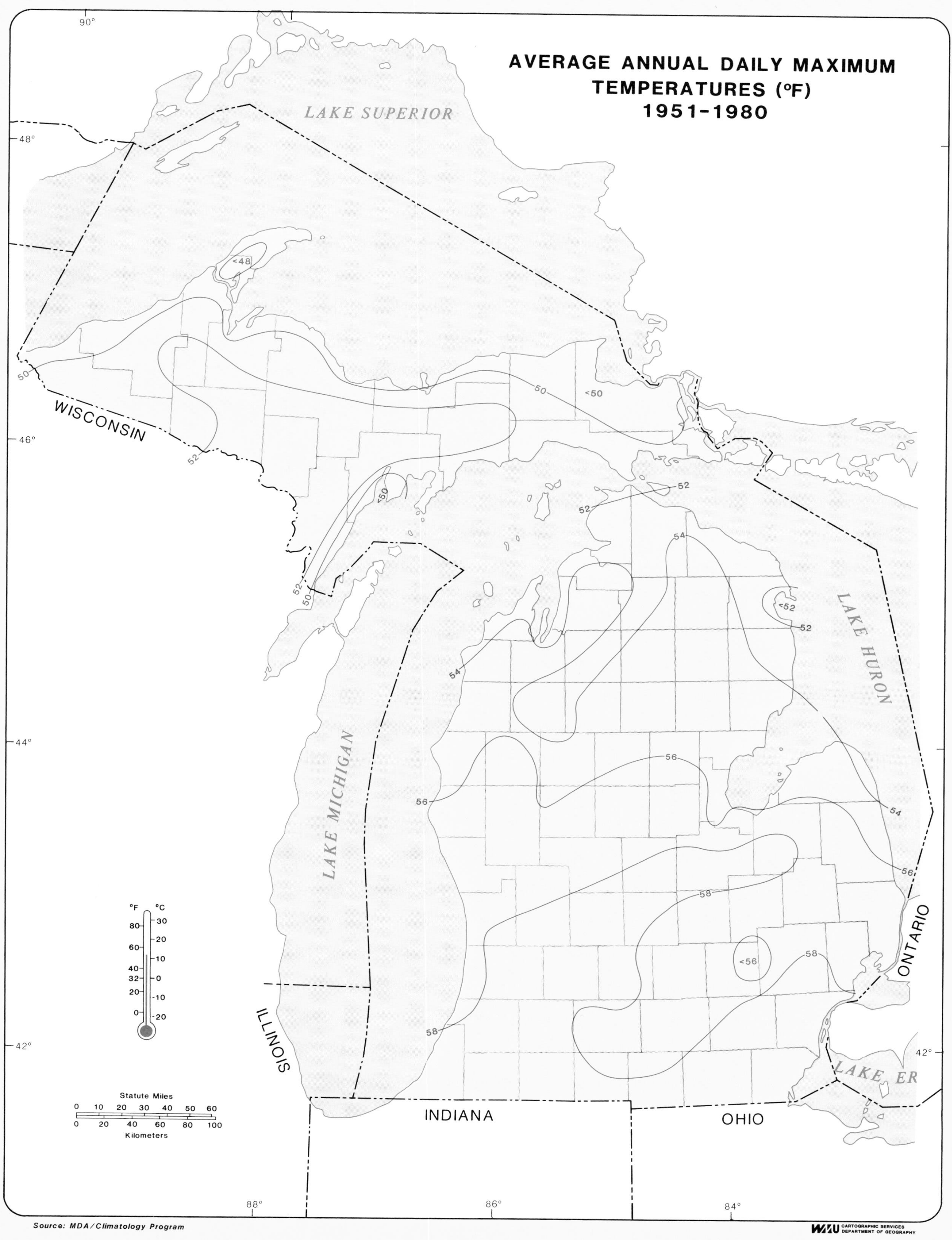


Source: MDA/Climatology Program

WMU Cartographic Services Department of Geography

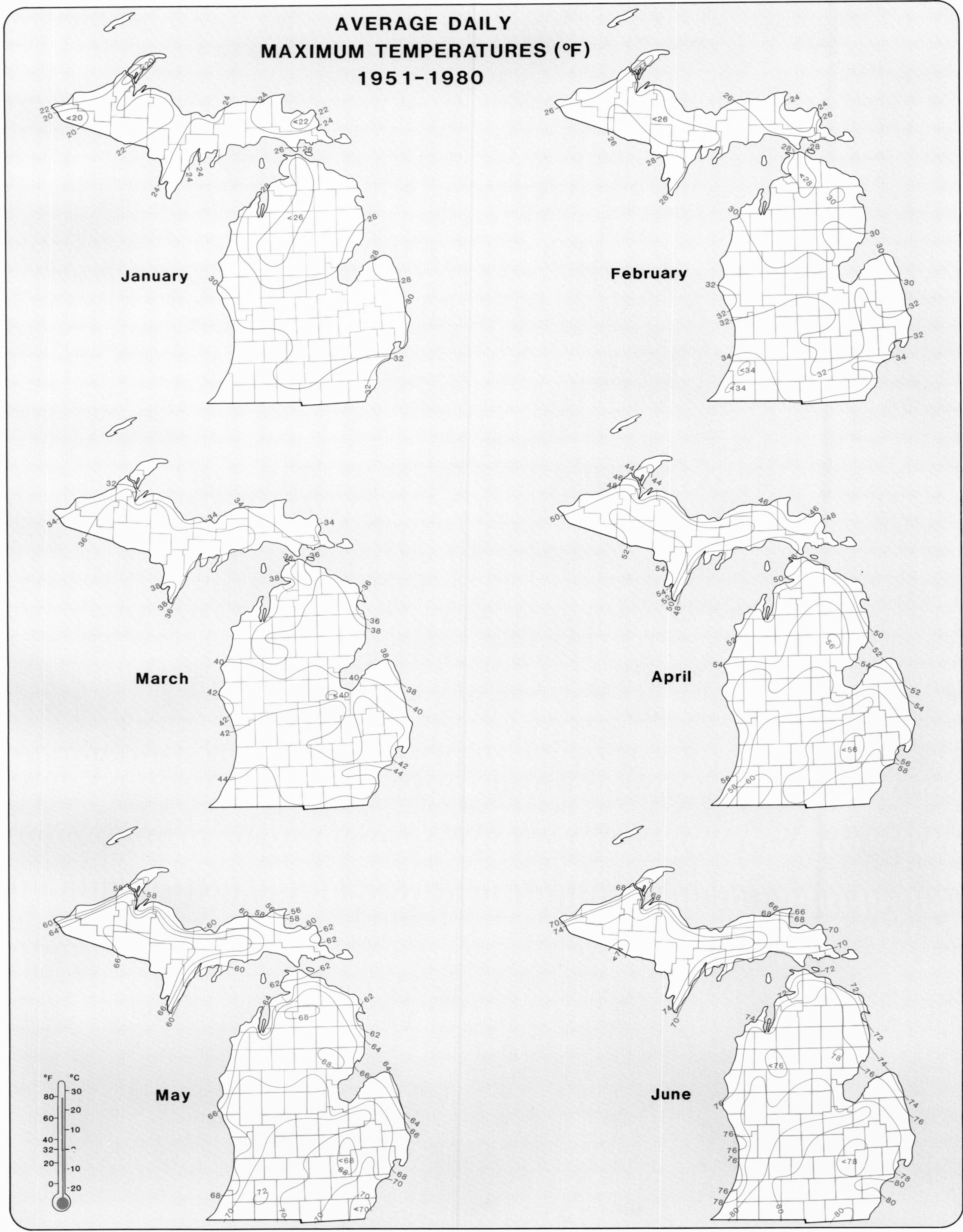


Source: MDA/Climatology Program

WMU CARTOGRAPHIC SERVICES DEPARTMENT OF GEOGRAPHY

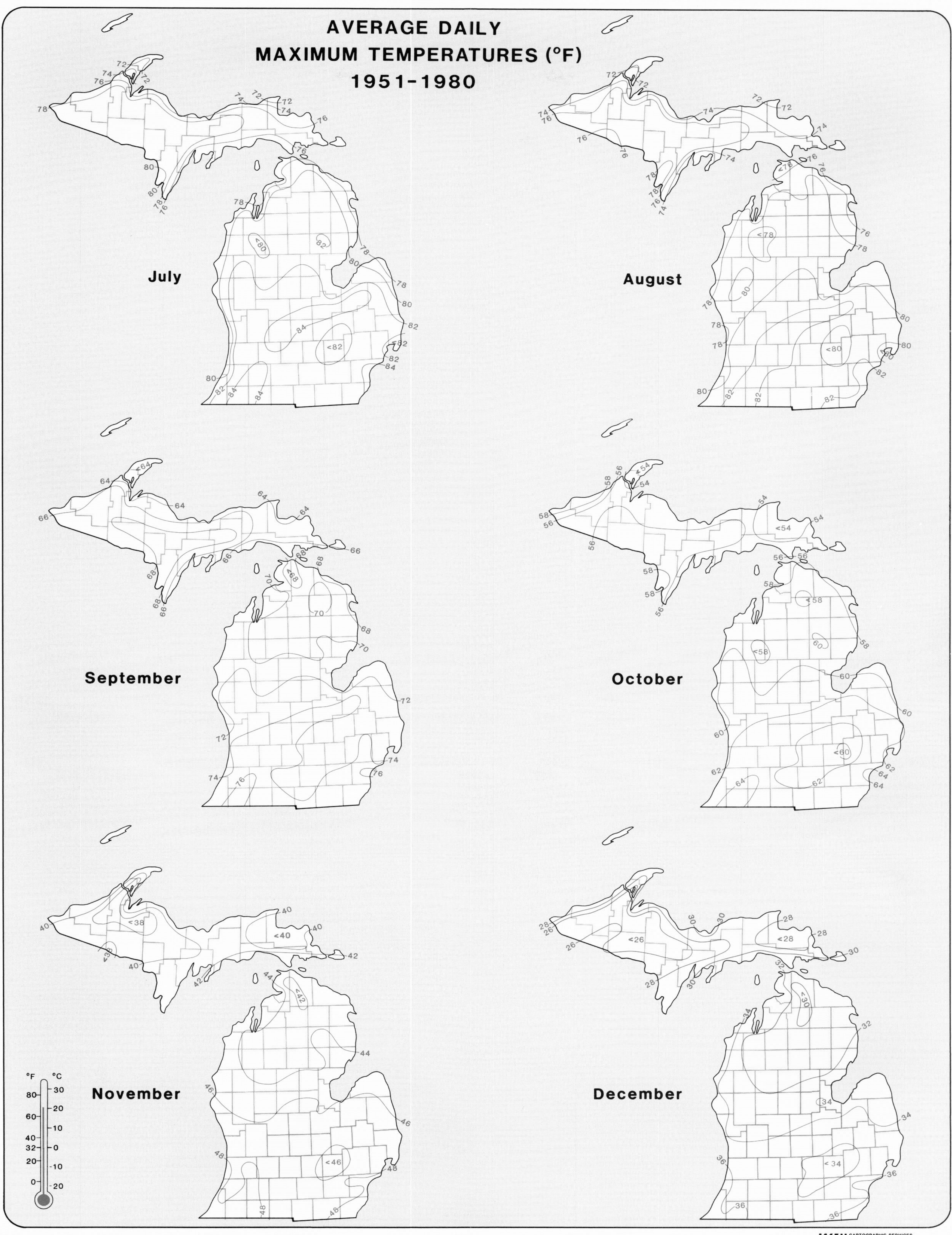


Source: MDA/Climatology Program

WMU CARTOGRAPHIC SERVICES DEPARTMENT OF GEOGRAPHY

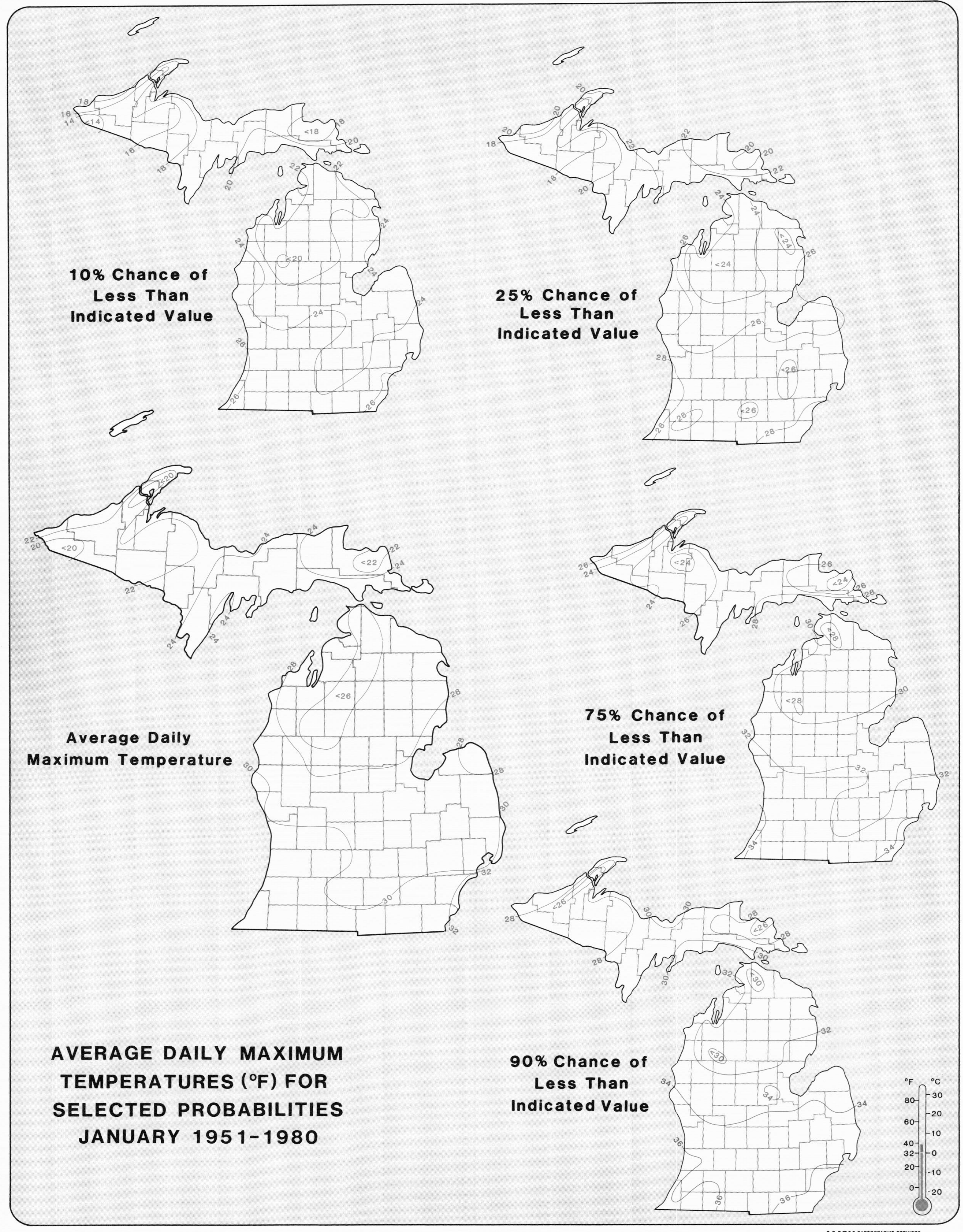


Source: MDA/Climatology Program

WMU CARTOGRAPHIC SERVICES DEPARTMENT OF GEOGRAPHY

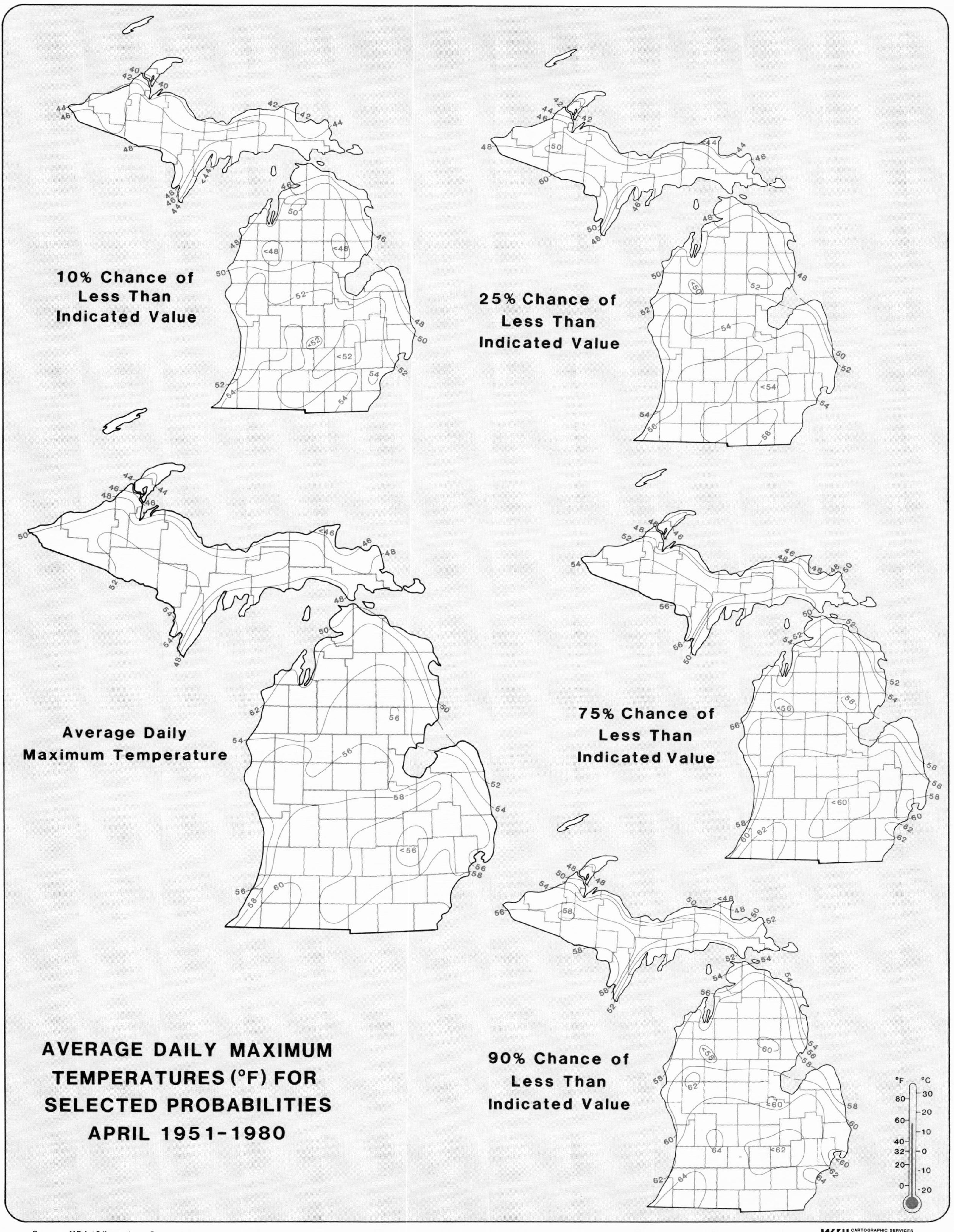
10% Chance of
Less Than
Indicated Value
25% Chance of
Less Than
Indicated Value
Average Daily
Maximum Temperature
75% Chance of
Less Than
Indicated Value
90% Chance of
Less Than
Indicated Value
AVERAGE DAILY MAXIMUM
TEMPERATURES (°F) FOR
SELECTED PROBABILITIES
APRIL 1951-1980
°F
°C
Source: MDA/Climatology Program
WMU CARTOGRAPHIC SERVICES
DEPARTMENT OF GEOGRAPHY

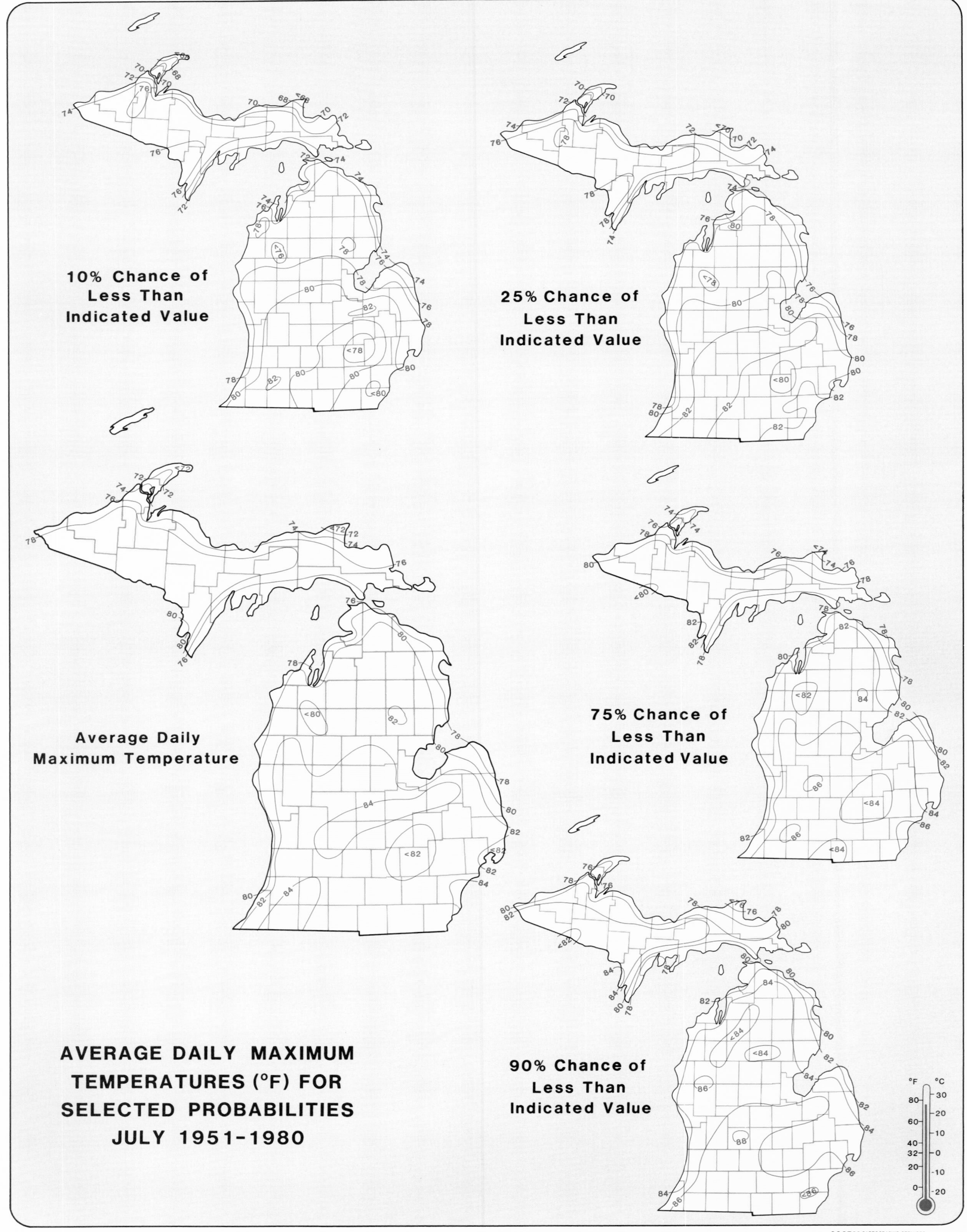


Source: MDA/Climatology Program

WMU CARTOGRAPHIC SERVICES DEPARTMENT OF GEOGRAPHY

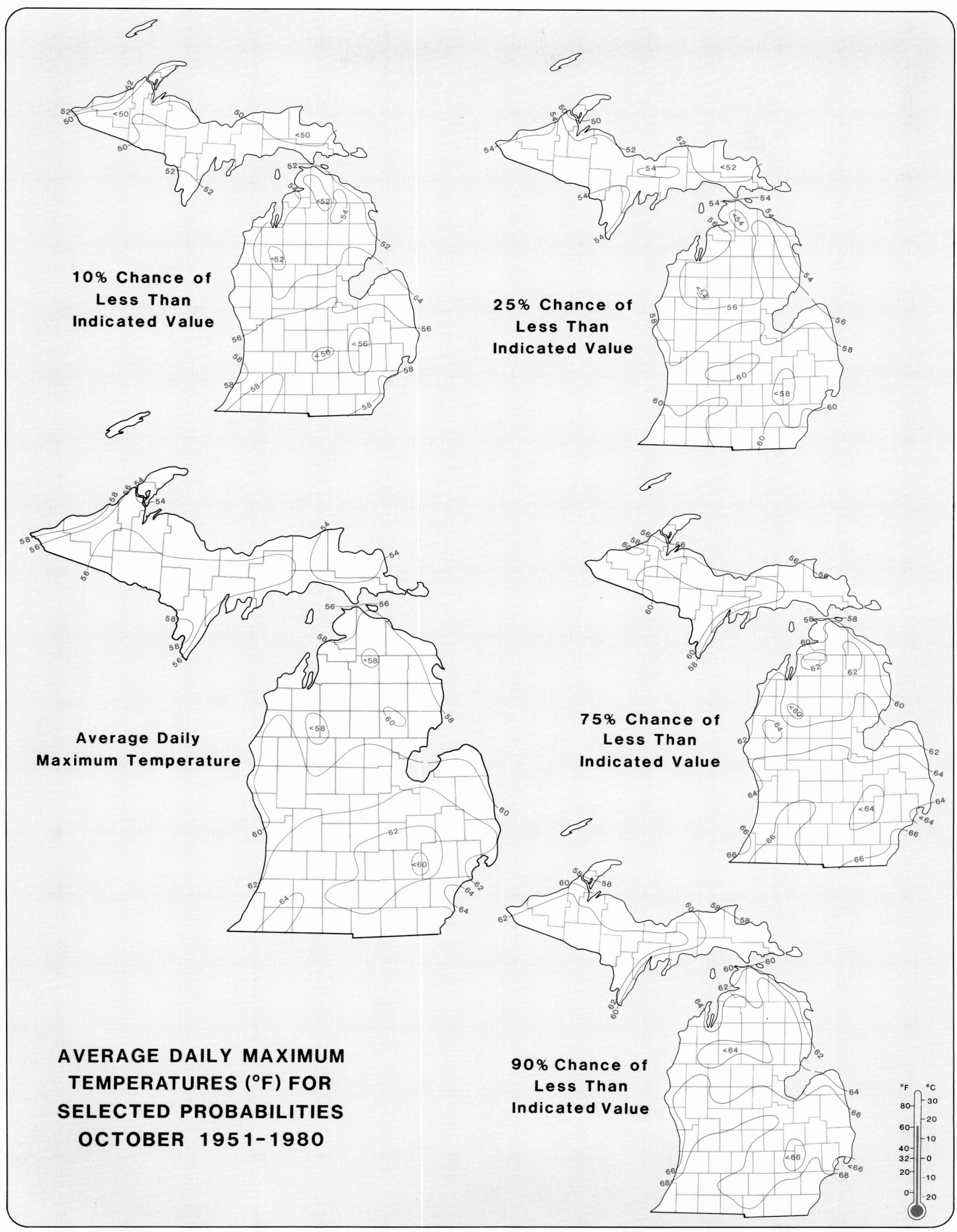


*Source: MDA/Climatology Program*

WMU CARTOGRAPHIC SERVICES DEPARTMENT OF GEOGRAPHY

# Average Number of Days with Maximum Temperatures of 90°F or Above

## *Seasonal* (*May–September*)

Ninety degree temperatures or higher have occurred at all stations in Michigan, although they are rare in the eastern part of the Upper Peninsula. The cooling effect of the Great Lakes, along with latitude, is a prominent control, with interior areas of the state experiencing more days 90°F or higher than lake shore areas. The largest number of days occurs in the southwestern and southeastern parts of the Lower Peninsula (Kalamazoo averages nearly 18 days, Dearborn nearly 17). However, at Fayette, on the north shore of Lake Michigan in the Upper Peninsula, 90°F or above temperatures may be expected only twice every ten years. At Whitefish Point, on the south shore of Lake Superior, only three out of ten years on the average will have a reading of 90°F or higher.

## *Monthly*

Readings of 90°F may occur early in the year as nine stations have recorded them in April during the 1951–1980 period. Interestingly, all readings were recorded in the northern Lower Peninsula and western Upper Peninsula.

There have been occasional occurrences of 90° temperatures at many Michigan stations in May, however no station within the state averages one or more days. The number recorded along the Lake Michigan shore north of Benton Harbor and in the eastern part of the Upper Peninsula is the smallest in the state. In June, many stations in the Lower Peninsula, except for immediate lakeshore areas, can expect at least one day with 90°F or higher. In the Upper Peninsula, only Marquette and Stephenson average one day or more in June. The largest average number of days occurs in southwestern and southeastern portions of the Lower Peninsula.

July averages the most occurrences, at least one day over the entire Lower Peninsula and the Upper Peninsula except for the Keweenaw Peninsula and portions of the east. The only station during the 30-year period to have recorded only one day with 90°F or greater in July is Fayette in the Upper Peninsula on the north shore of Lake Michigan. Whitefish Point, on the south shore of Lake Superior, can expect a 90° day in July only once in ten years. On the other hand, Kalamazoo and Dearborn average about 7 days of 90°F or higher temperatures in July.

Occurrences of 90°F or greater decrease in August, but all stations have experienced at least one occurrence over the years. By September, only the southern portion of the Lower Peninsula can expect at least one 90° day, while two stations in the eastern Upper Peninsula, Fayette and Whitefish Point, did not record a 90°F or higher day during the 30-year period.

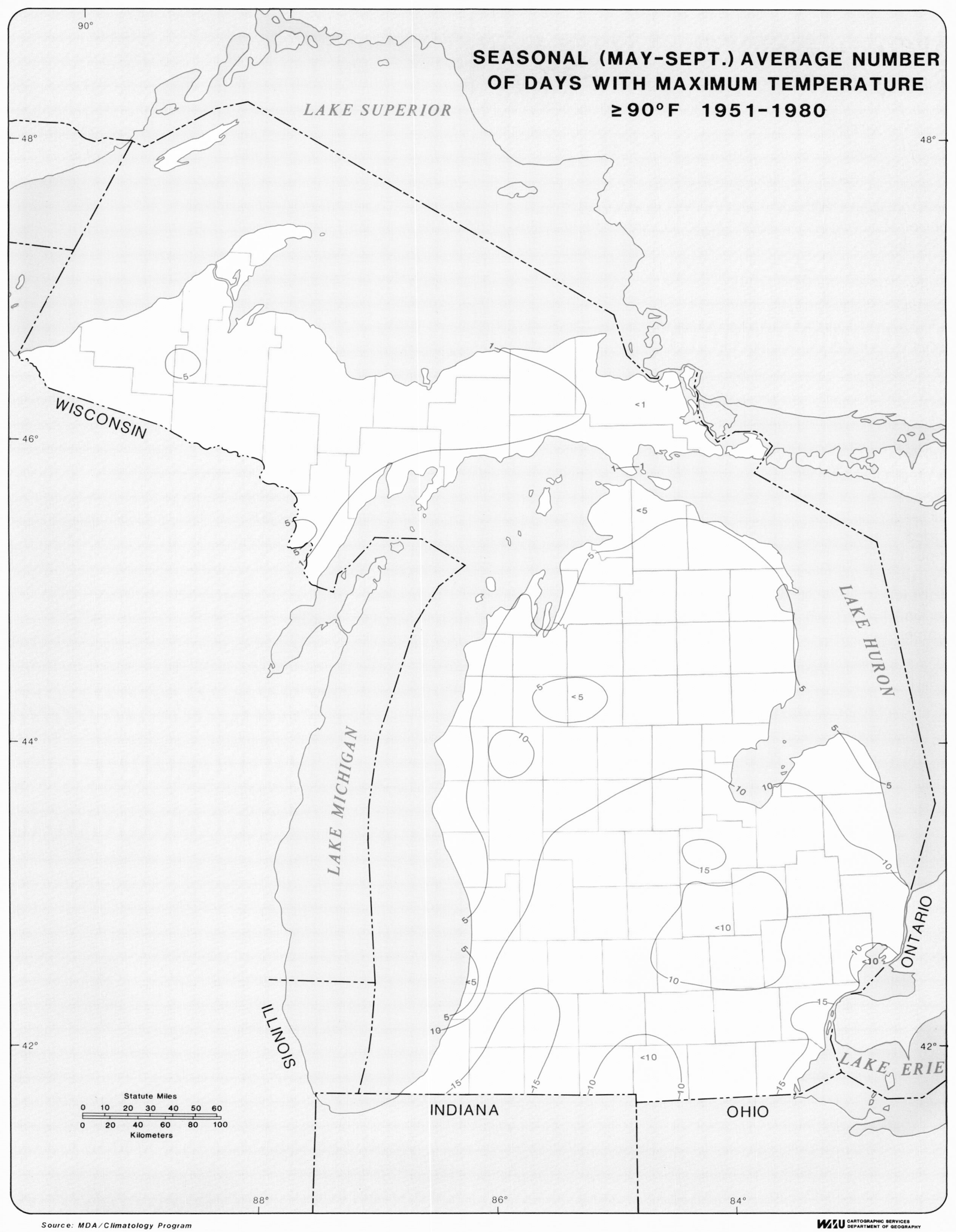


Source: MDA/Climatology Program

WMU CARTOGRAPHIC SERVICES DEPARTMENT OF GEOGRAPHY

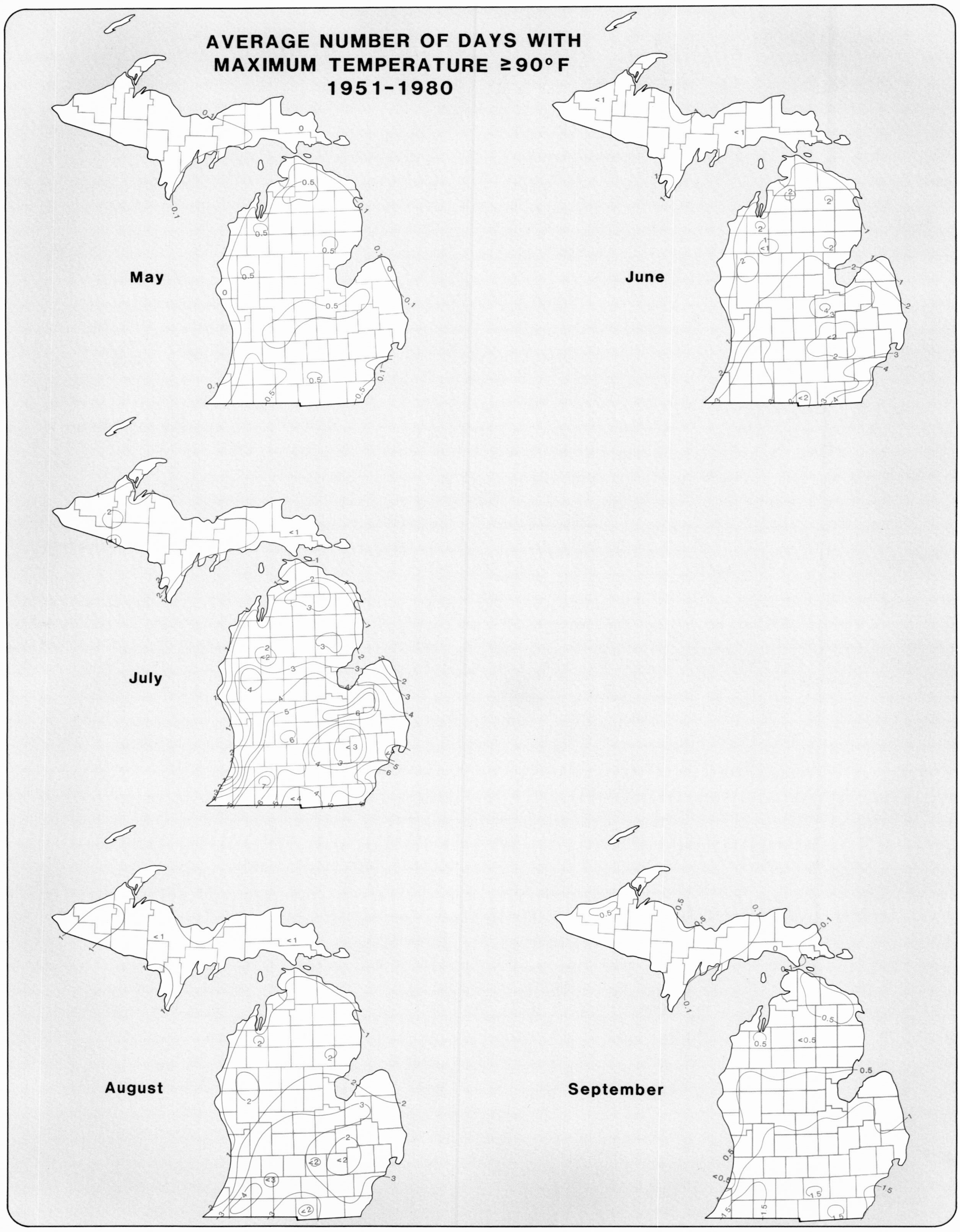


Source: MDA/Climatology Program

WMU CARTOGRAPHIC SERVICES DEPARTMENT OF GEOGRAPHY

# Average Number of Days with Maximum Temperature 32°F or Below

## *Seasonal* (*October–May*)

During the colder time of the year there are many days when maximum temperatures fail to rise above the freezing point (freezing days). On the Keweenaw Peninsula the Houghton FAA Airport station leads the state with an average of 109 days (or nearly 30% of the days during the year). In the Detroit metropolitan area, Grosse Pointe Farms averages only 42 freezing days, the least number for any station in Michigan during the 30-year period.

Although latitude is an important control in determining whether the daily maximum will rise above the freezing point, the lakes also affect the seasonal and monthly averages. Lakeshore regions normally experience fewer days, due to heat release to the nearby shores on days when temperatures otherwise might remain slightly below 32°F.

## *Monthly*

Although October is the first month of the cool season when maximum temperatures of 32°F or less may occasionally occur, most stations in Michigan do not experience such a day. In the northern interior portion of the Lower Peninsula, a few stations may experience a freezing day in October. In the Upper Peninsula such days are more common, with Houghton FAA Airport expecting 4 years out of 10 with one day having temperatures failing to rise above freezing.

In November, averages range from less than 2 days in the extreme southwest and southeast areas of the Lower Peninsula to over 10 at Houghton. By December, with winter in full sway, an average of 24 (or 77% of the days) have maximum temperatures of freezing or below at Houghton, while at South Haven, along the Lake Michigan shore, the average is only 10 days.

In January, the coldest month, over half the days at all Michigan stations fail to see temperatures above freezing. The smallest average number of freezing days is 16 at Monroe and South Haven, while at Houghton the average is 28, or 90% of the days.

February brings some decrease in the number of freezing days although Houghton still averages 24, or 86% of the days. In the southern Lower Peninsula, with the slowly advancing warmth of February, more than half of the daytime maxima rise above 32°F. By March, in the extreme south, less than 4 days fail to rise above 32°F, but freezing days are still common in the Upper Peninsula, with an average of 18 at Houghton.

April brings much warmer temperatures with a number of stations in the southern part of the Lower Peninsula not having recorded a freezing day during the 30-year period. Pellston is the only station in the Lower Peninsula that can expect one freezing day in April, although Houghton in the Upper Peninsula averages 3.

In May, most stations in the state did not record a freezing day during the 30-year period although several stations in the western Upper Peninsula did experience freezing days.

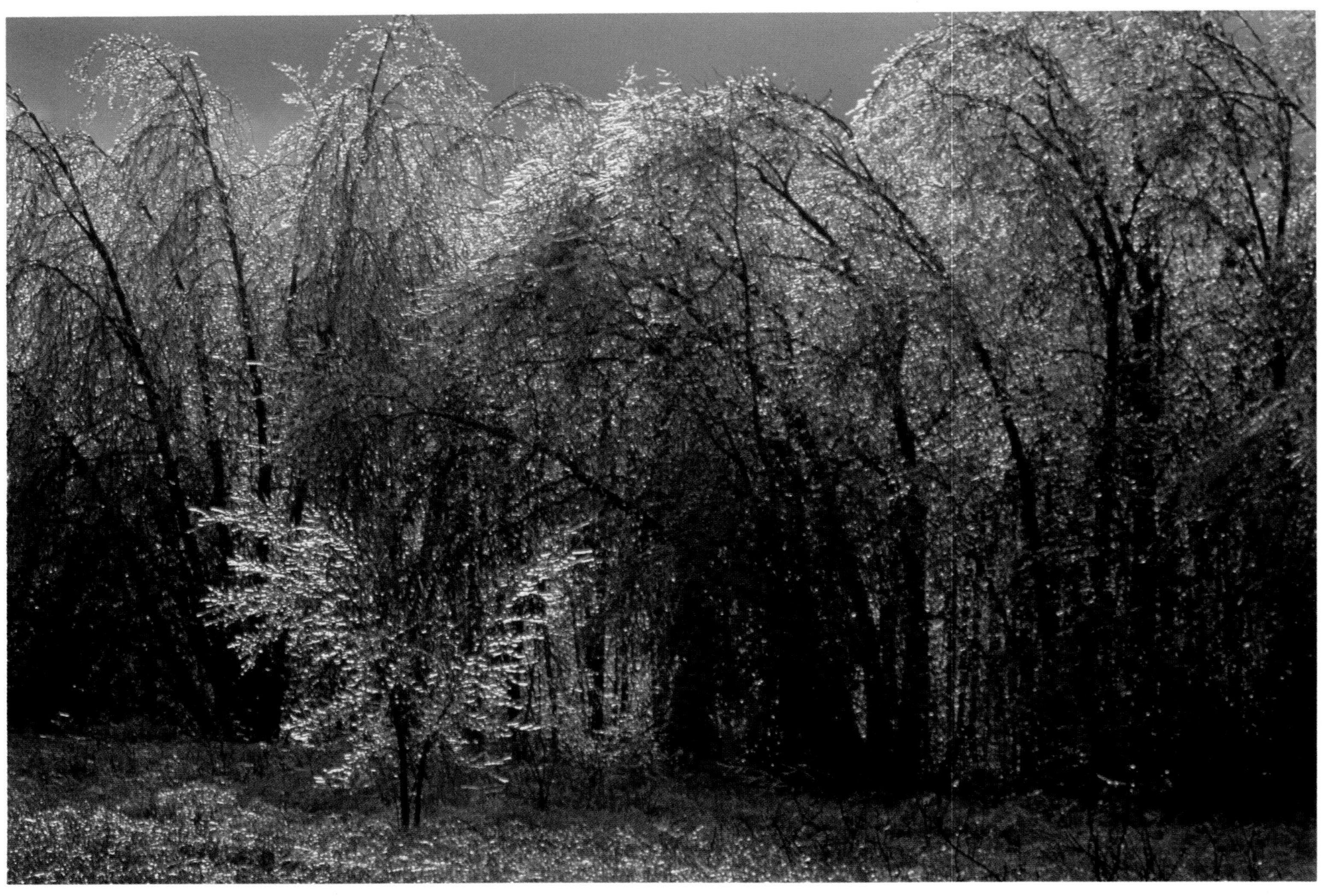

New Year's Eve icestorm, Kalamazoo, 1984.

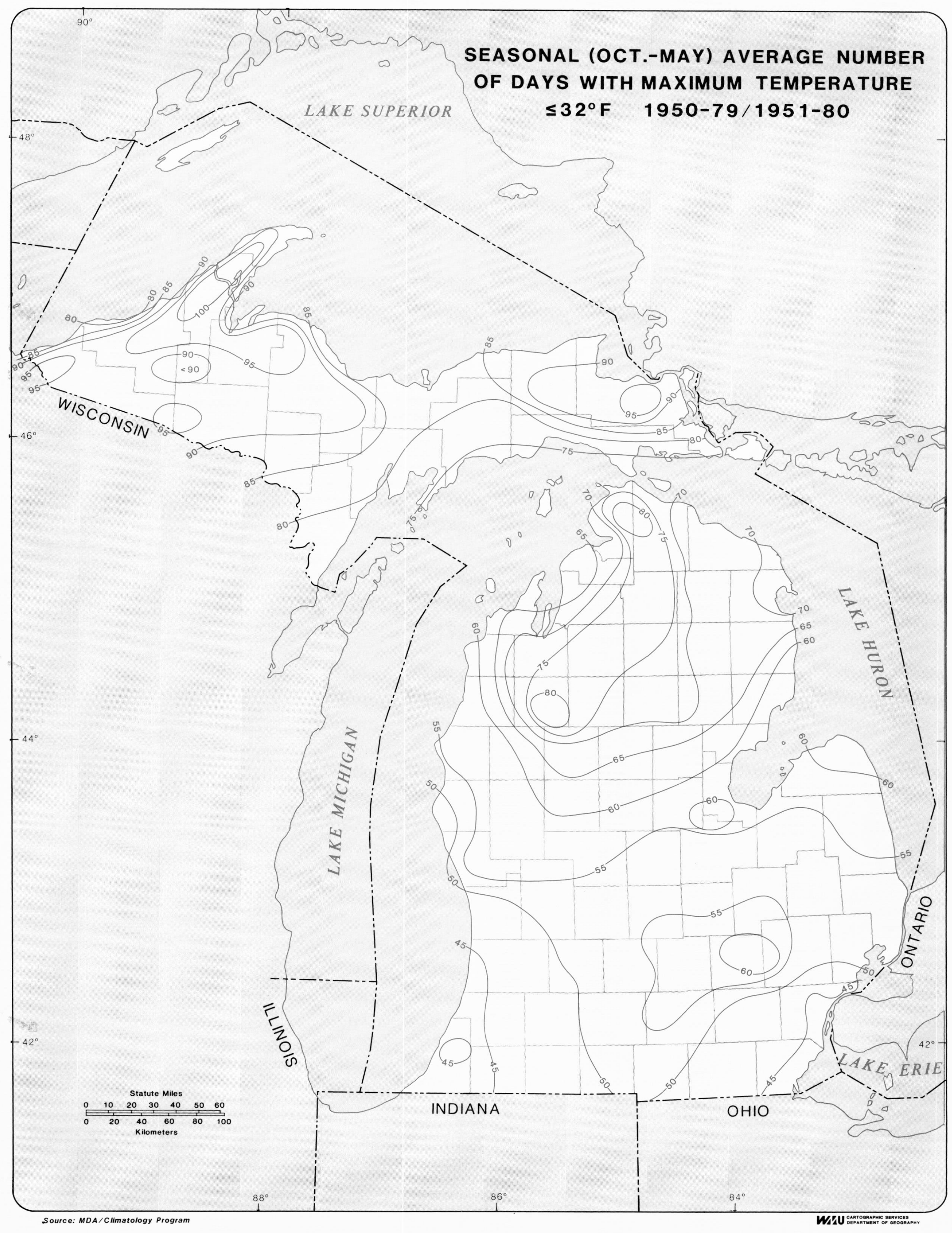


Source: MDA/Climatology Program

WMU Cartographic Services, Department of Geography

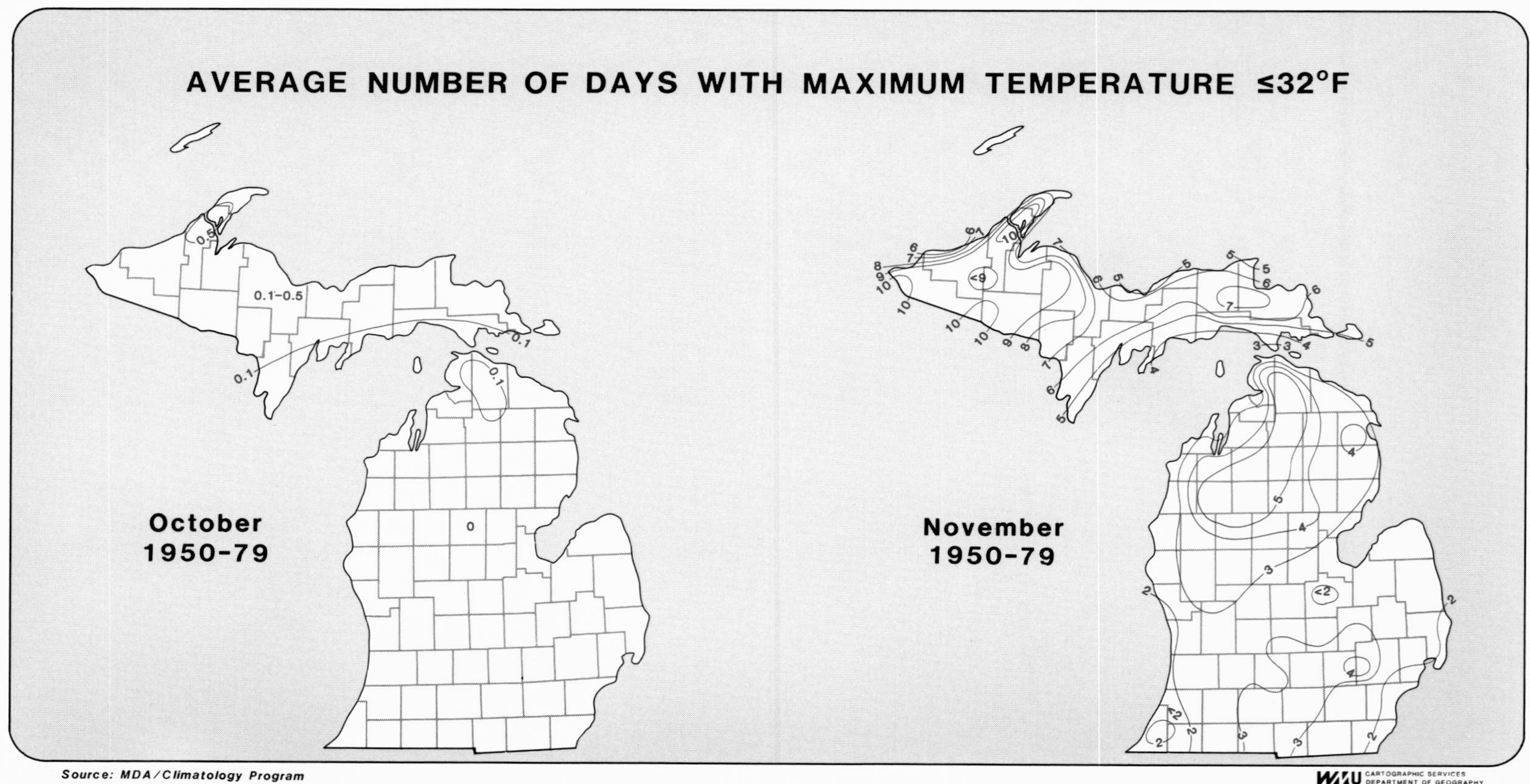
AVERAGE NUMBER OF DAYS WITH MAXIMUM TEMPERATURE ≤32°F
October
1950-79
November
1950-79
Source: MDA/Climatology Program
WMU CARTOGRAPHIC SERVICES DEPARTMENT OF GEOGRAPHY

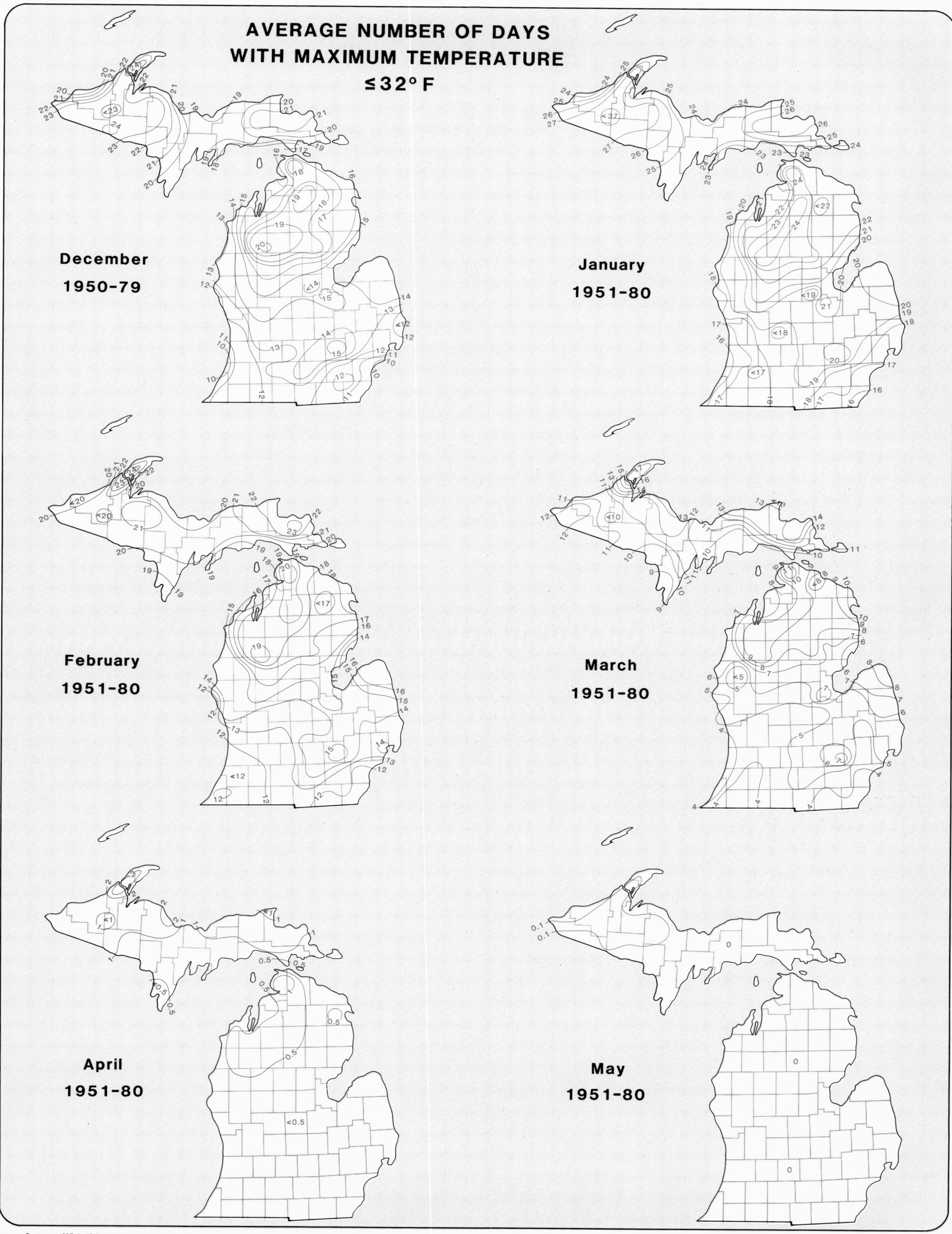


*Source: MDA/Climatology Program*

WMU CARTOGRAPHIC SERVICES DEPARTMENT OF GEOGRAPHY

# Average Daily Minimum Temperatures

## *Average Annual Daily Minimum Temperatures*

Lake effects play a strong role in determining annual average daily minimum temperatures (normally the nighttime lows). Interior areas, especially in the winter, are likely to have lower nighttime temperatures than areas near the shores of the lakes. In Michigan, the coldest nights, averaged for the year, occur in the interior of the western Upper Peninsula. Bergland Dam has the lowest average value, 27.3°F. Average annual daily minimum temperatures are below freezing over a large portion of the Upper Peninsula and the interior of the northern Lower Peninsula.

The warmest daily minima occur within the Detroit metropolitan area at the Detroit City Airport, with an annual average of 41.8°F. In this southern portion of the state, the heat island effect, usually most noticeable at night, combines with latitude to produce comparatively warm nights. The contrast in average annual minimum temperatures between Bergland Dam and Detroit City Airport is 14.5°F.

## *Average Daily Minimum Temperatures (Monthly)*

During all months of the year the warmest average daily minimum temperatures occur at Detroit City Airport. In July, the Detroit City Airport experiences the highest average daily minimum temperatures of any Michigan station, 64.1°F. In January, the average daily minimum at Detroit City Airport is, of course, much colder (18.8°F), but still the warmest in Michigan for that month. High nocturnal minima also occur during all months of the year along the shores of Lake Michigan in the southwestern portion of the Lower Peninsula, where heat stored by Lake Michigan prevents low nighttime temperatures from occurring.

The coldest nights during the winter months, December through March, occur at Bergland Dam, in the western interior portion of the Upper Peninsula. In January, Bergland Dam experiences an average daily minimum temperature of −0.6°F. Champion at 0.4°F, and Ironwood and Kenton at 0.8°F are close behind. The effects of the lakes in elevating minimum temperatures are prominent during the winter and isotherms are packed along the lake shores. In April and May, the core area of low minima shifts eastward in the Upper Peninsula to Champion, and in June to Whitefish Point. In July and August, the coolest nights occur at Vanderbilt in the interior of the northern Lower Peninsula (July average daily minimum 49.6°F). From September through November, the cold axis shifts back westward in the Upper Peninsula, to Champion, and then in December it returns to Bergland. From November through April, daily minimum temperatures average below freezing somewhere within the state.

The range between coldest and warmest nighttime temperatures is largest in February (20.1°F) and least in October (11.9°F). Little warmup in nocturnal monthly average temperatures is evident in February as less cloudiness and more extensive ice cover on the lakes allow nighttime temperatures to plunge to values that are nearly as cold as those in January.

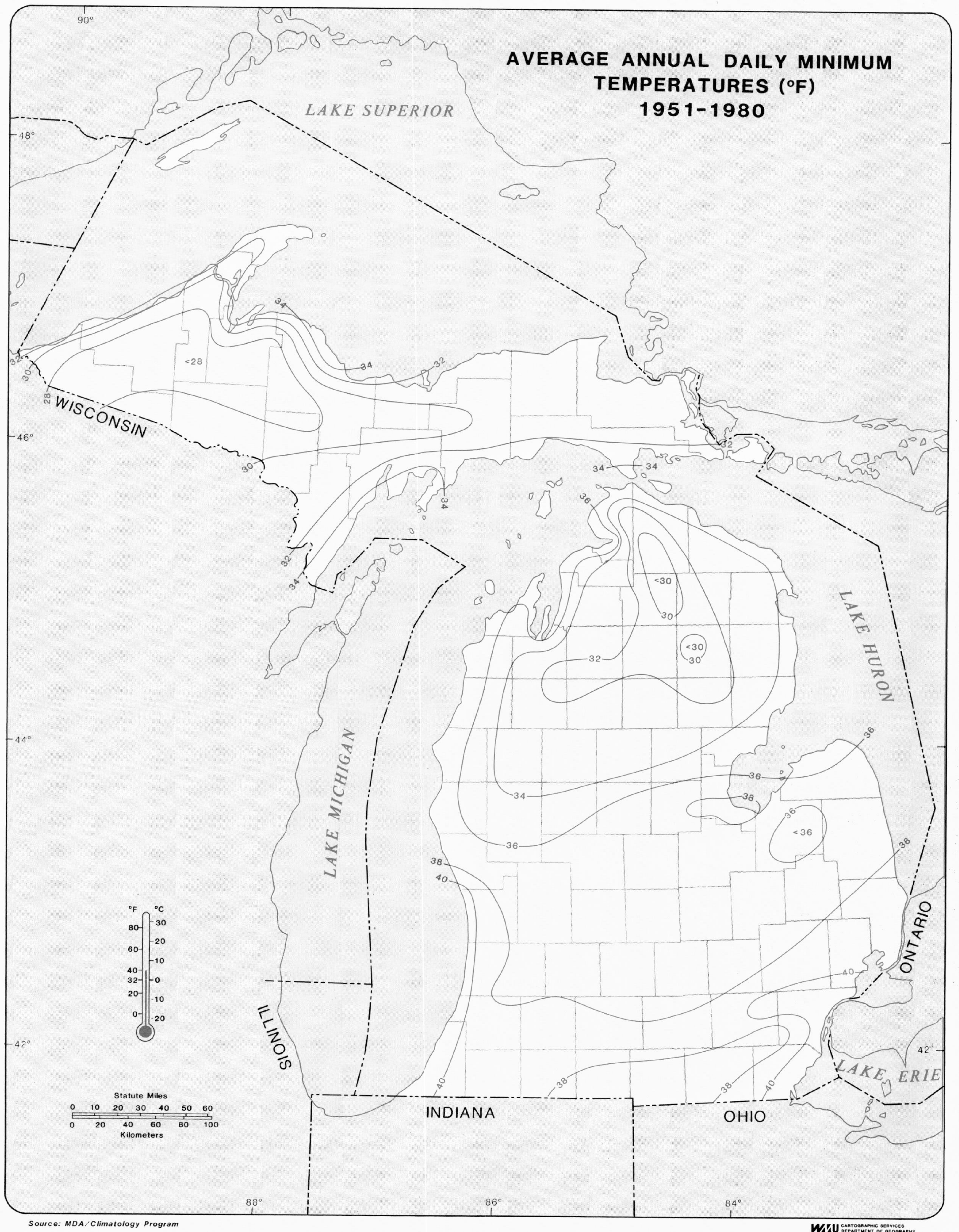


Source: MDA/Climatology Program

WMU CARTOGRAPHIC SERVICES DEPARTMENT OF GEOGRAPHY

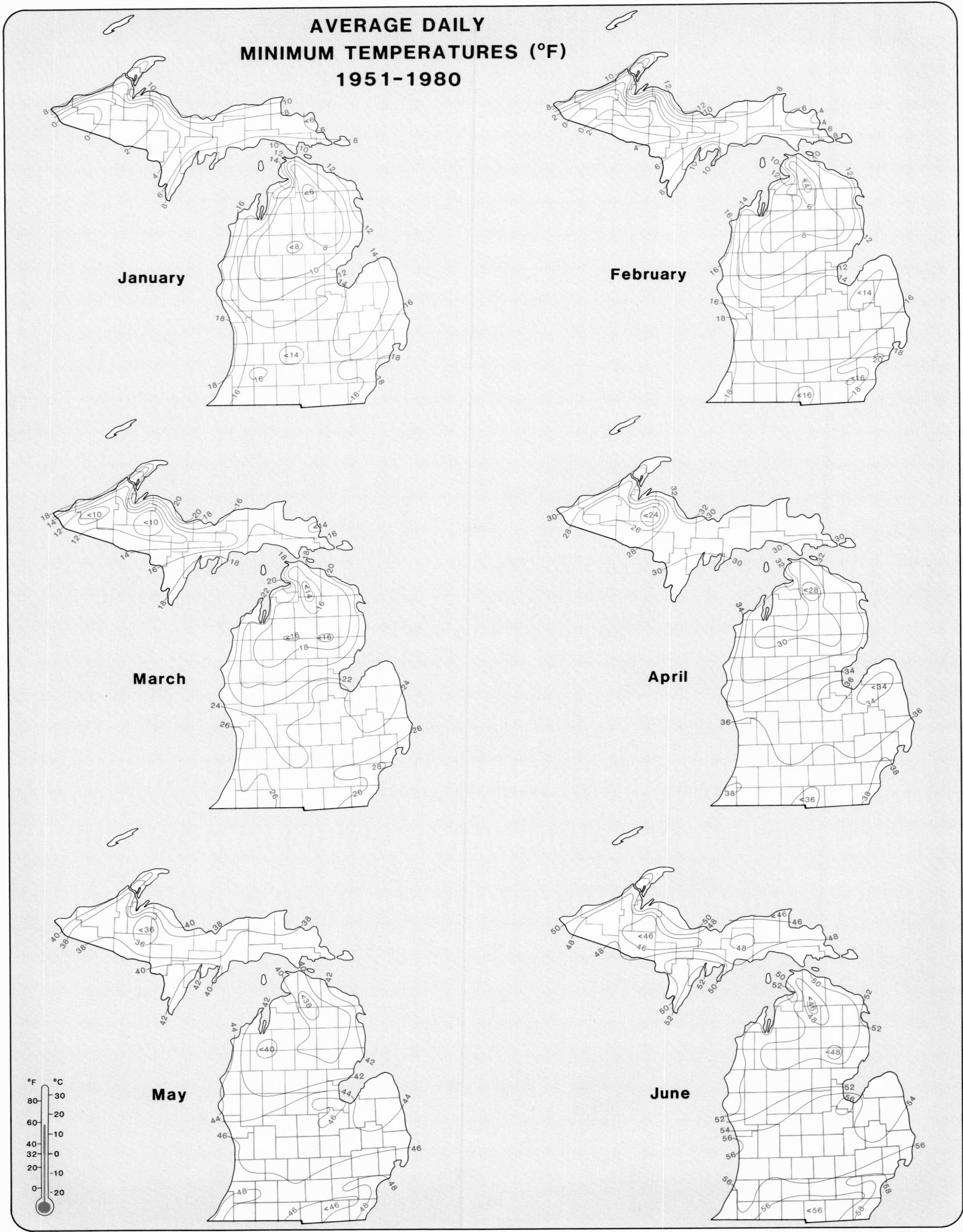
AVERAGE DAILY
MINIMUM TEMPERATURES (°F)
1951-1980
January
February
March
April
May
June
°F
°C
Source: MDA/Climatology Program
WMU CARTOGRAPHIC SERVICES DEPARTMENT OF GEOGRAPHY

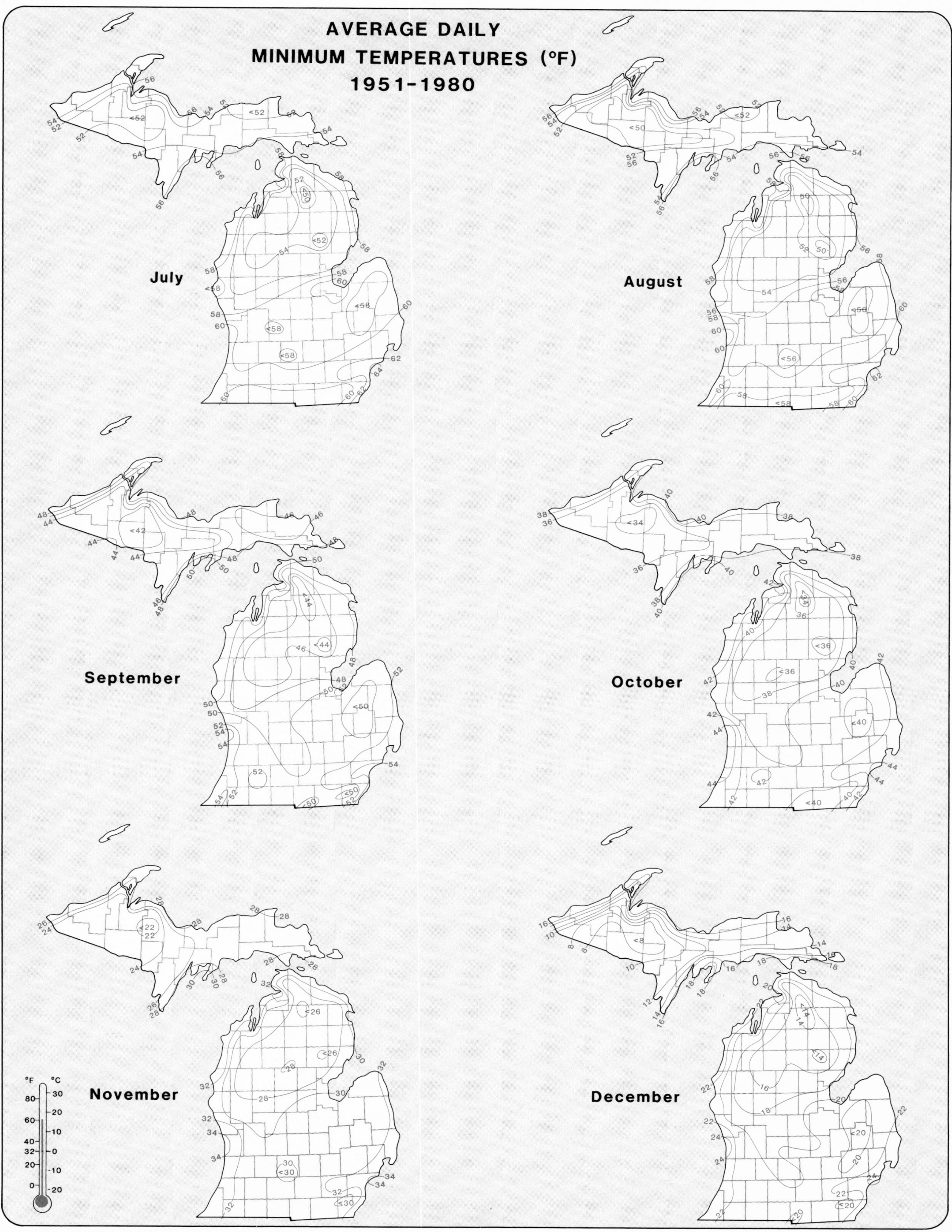


Source: MDA/Climatology Program

WMU CARTOGRAPHIC SERVICES DEPARTMENT OF GEOGRAPHY

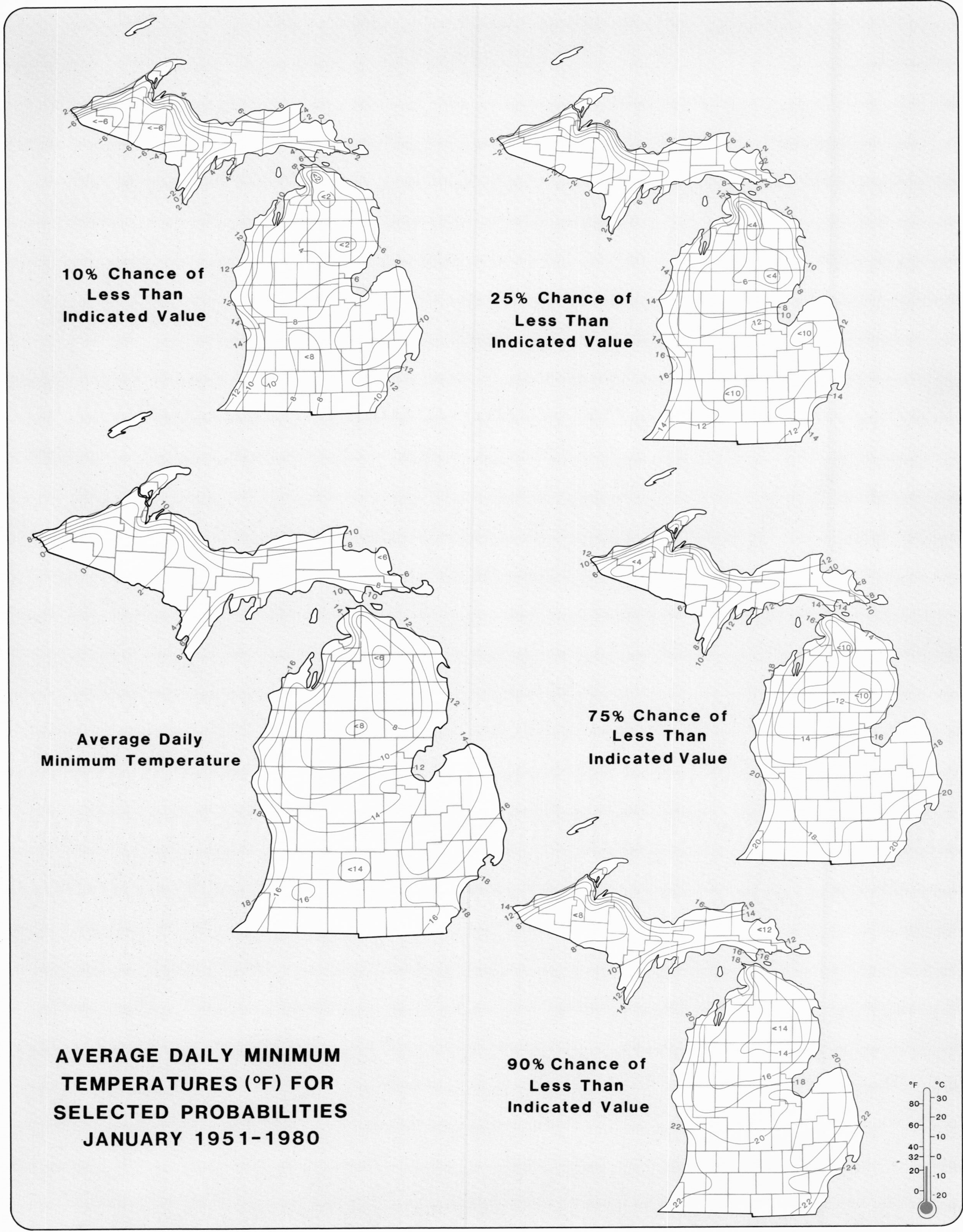


Source: MDA/Climatology Program

WMU CARTOGRAPHIC SERVICES DEPARTMENT OF GEOGRAPHY

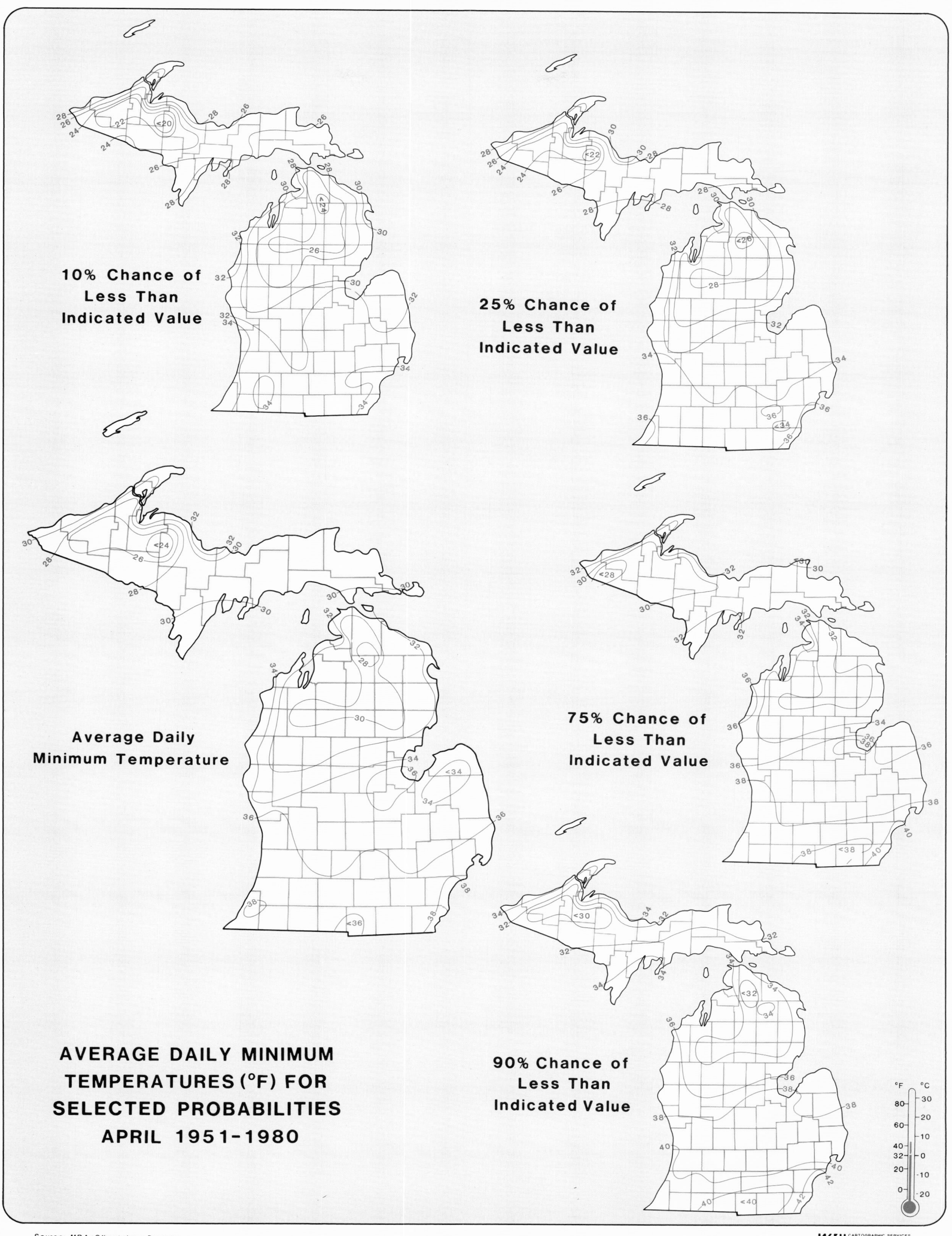


Source: MDA/Climatology Program

WMU CARTOGRAPHIC SERVICES DEPARTMENT OF GEOGRAPHY

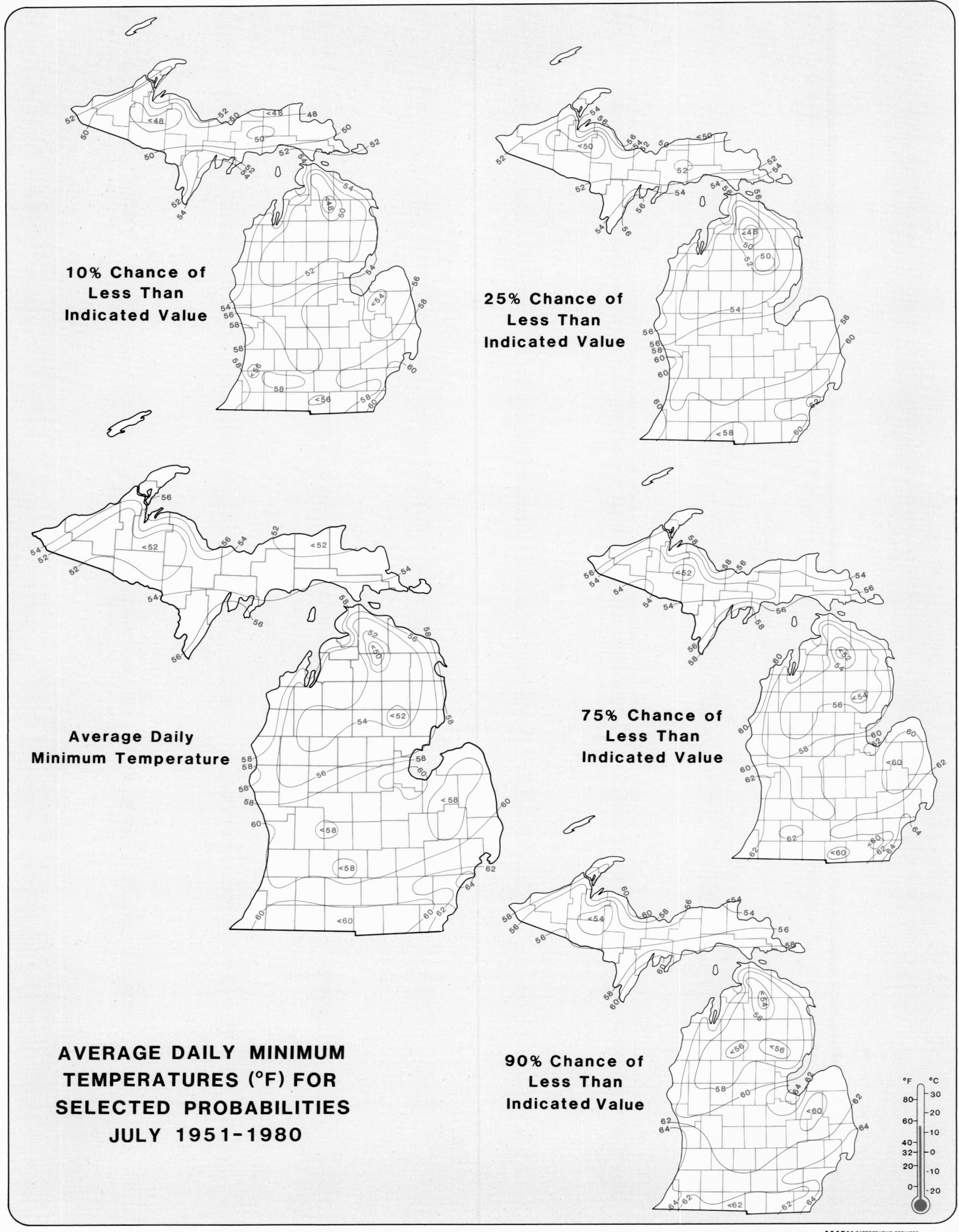


Source: MDA/Climatology Program

WMU CARTOGRAPHIC SERVICES DEPARTMENT OF GEOGRAPHY

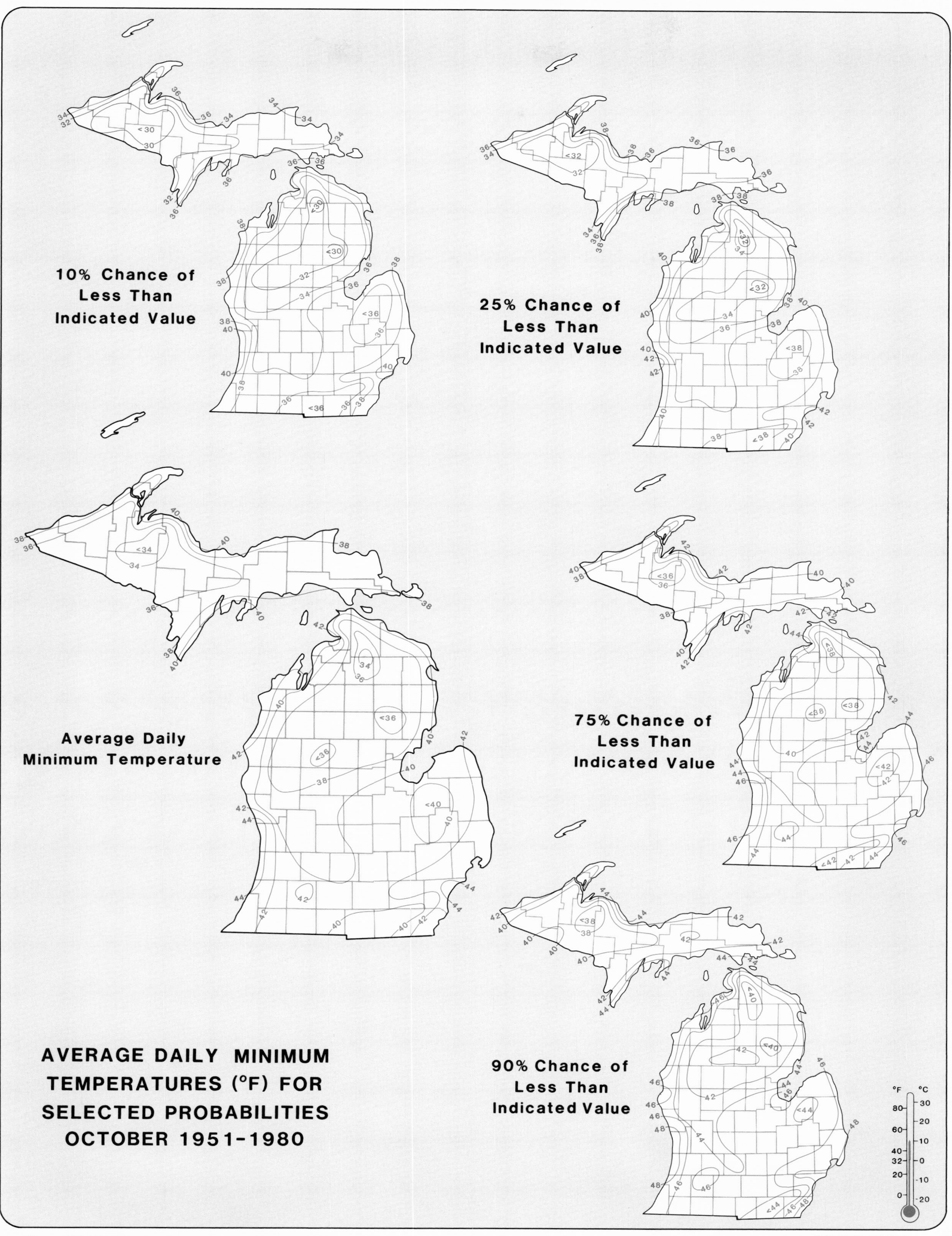


Source: MDA/Climatology Program

WMU CARTOGRAPHIC SERVICES DEPARTMENT OF GEOGRAPHY

# Average Number of Days with Minimum Temperatures 32°F or Below

## *Seasonal (July–June)*

Freezing nighttime temperatures are the rule rather than the exception in Michigan. At least one station within the state can expect a freeze night during all months of the year except July. The areas experiencing the largest number of freeze nights include the western interior of the Upper Peninsula and the northern interior of the Lower Peninsula. The Champion station in the western interior portion of the Upper Peninsula averages 210 or about 58% of days during the year with minimum temperatures reaching freezing or below. The least number of days occurs in the Detroit metropolitan area and near the shore of Lake Michigan in the southwestern portion of the Lower Peninsula. At Detroit City Airport an average of only 122 or 33% of the days have minima of 32°F or less.

## *Monthly*

In August, normally one night with freezing temperatures can be expected at Champion and Stambaugh and two at Vanderbilt. By September, much of the Upper Peninsula except immediate lakeshore areas, and a large part of the northern Lower Peninsula can expect some freeze nights. By October, all stations in the state can expect freeze nights, ranging from an average of about half the nights at Vanderbilt, Bergland Dam, Champion, and Kenton to only two nights at Detroit City Airport and Grand Haven. In November, the frequency increases to 26 at stations in the western Upper Peninsula, and 13 at Detroit City Airport and Grosse Pointe Farms. From December through February, nearly all nights at all stations in the state can expect freezing temperatures, while in March 26 to 30 days in the north and 21 to 25 days in the south have average minima of 32°F or less. April is a transition month. All stations can expect some freezing nights, but the range is large; from Champion and Bergland Dam's 24 to Detroit City Airport's 7. In May, although all stations can expect at least one freeze night, the majority of the minima are above freezing at all stations. In June, only the interior of the Upper Peninsula (Kenton and Champion, 3), the Lake Superior shoreline between Marquette and Munising, and the interior of the northern Lower Peninsula (Vanderbilt, 4) can expect nighttime temperature minima of freezing or below.

Source: MDA/Climatology Program

WMU CARTOGRAPHIC SERVICES DEPARTMENT OF GEOGRAPHY

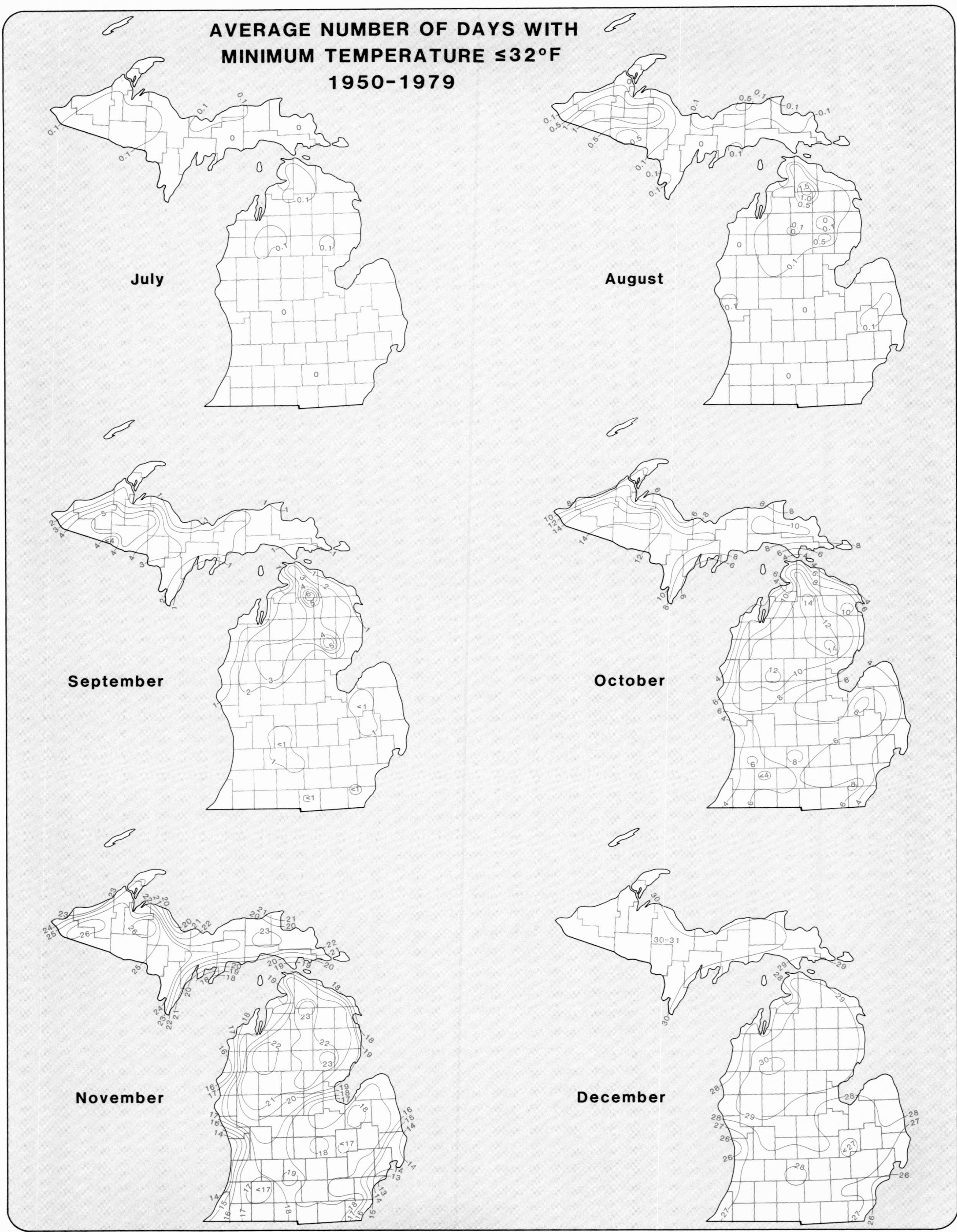


*Source: MDA/Climatology Program*

WMU CARTOGRAPHIC SERVICES DEPARTMENT OF GEOGRAPHY

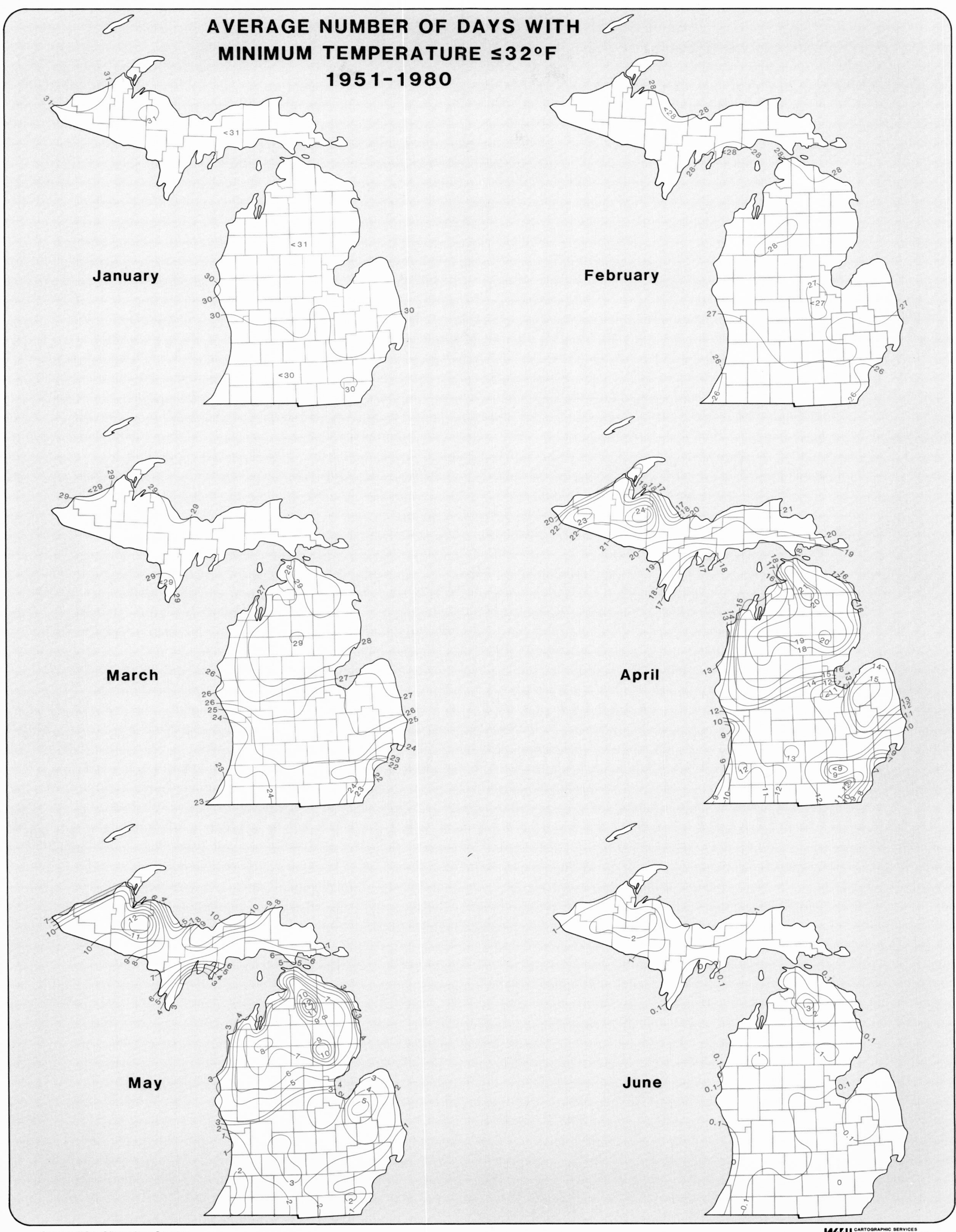


Source: MDA/Climatology Program

WMU CARTOGRAPHIC SERVICES DEPARTMENT OF GEOGRAPHY

# Average Number of Days with Minimum Temperatures 0°F or Below

## *Seasonal* (*November–April*)

Temperatures of zero or below normally occur during the winter at all stations in Michigan, although there is a large variation within the state. Latitude and, in particular, lake proximity are the major controls. The interior area of the western Upper Peninsula, and the northern interior of the Lower Peninsula average the most days. Bergland Dam has an average of 51 days (or 28% of the days from November-April) with zero or below temperatures.

On the other hand, Grand Haven, warmed by Lake Michigan, and Detroit City Airport within the Detroit urban heat island, average only 2 days with zero or below temperatures. Other stations in the southern part of the state normally experience only 5–10 days per year.

## *Monthly*

Zero or below temperatures begin to occur in the western interior of the Upper Peninsula in November, although only one to two days can be expected. By December, all stations in the state except those along the Lake Michigan shore and in the extreme southeastern part of the Lower Peninsula average at least one occurrence of zero or below temperatures. In the interior of the western Upper Peninsula between 8 and 9 days are experienced.

January has the maximum number of days in all areas, ranging from nearly 16 at Bergland Dam to 1 at Grand Haven, South Haven, Muskegon, and Detroit City Airport. February, with less cloudiness and less warmth from the lakes, still experiences many zero or below readings, in spite of the fact that it is a short month. Over half the days at Bergland Dam (15) experience zero or below, while a number of stations in southwestern and southeastern Lower Michigan average only one day.

March shows a marked warming in the southern part of the state where most stations normally do not average one day with a zero or below reading. Detroit City Airport is the only station in the 1951–1980 period that did not record a zero or below temperature in March. On the other hand, occurrences are still common in the Upper Peninsula, with Bergland and Champion (9 days) having nearly one-third of the nights with zero or below temperature.

Most stations in the southern and central portions of the Lower Peninsula have not experienced zero or below readings in April. They may occur during some years in the Upper Peninsula and northern Lower Peninsula. Only Champion, Alberta, Bergland Dam, Kenton, and Watersmeet average one occurrence in April.

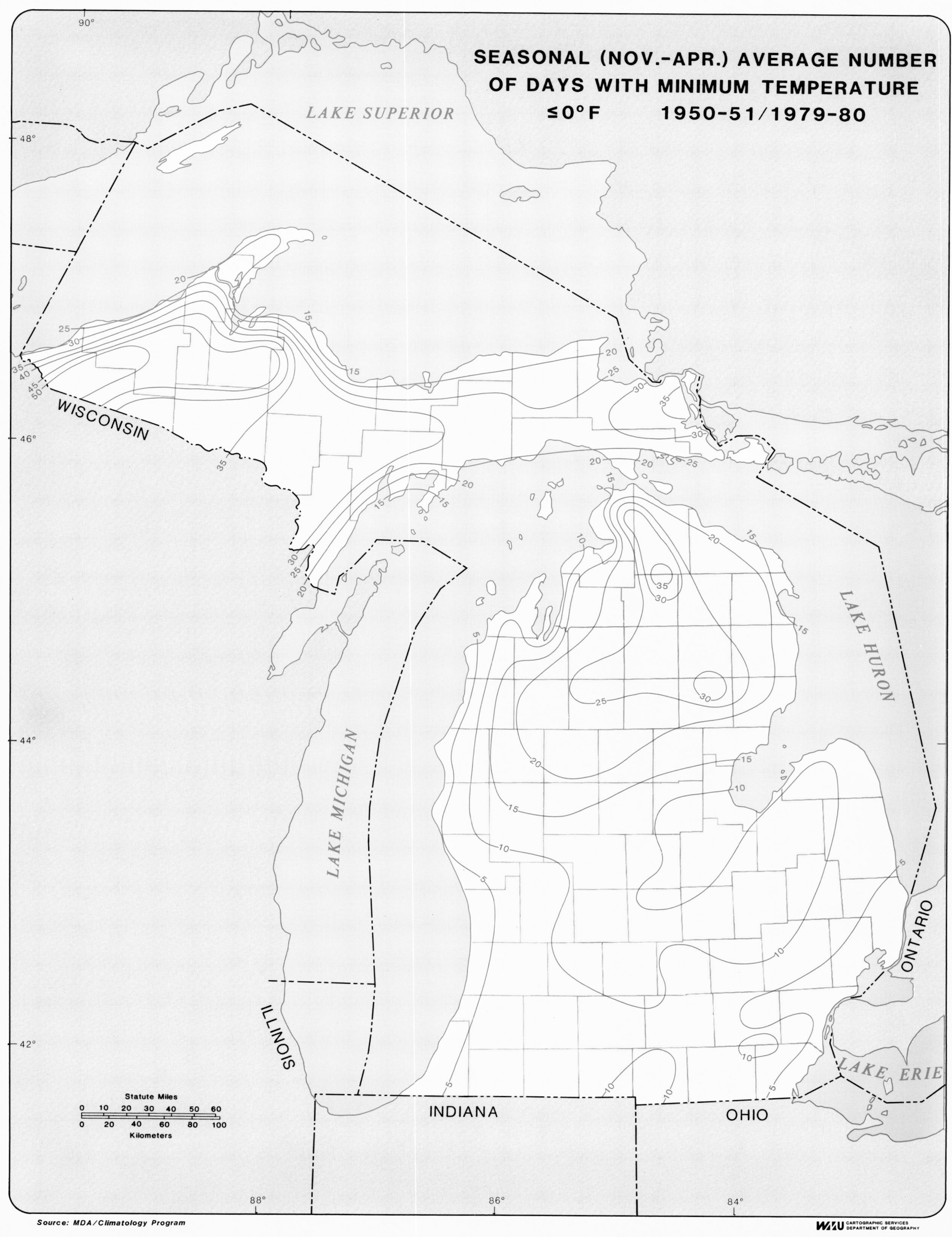


Source: MDA/Climatology Program

WMU CARTOGRAPHIC SERVICES DEPARTMENT OF GEOGRAPHY

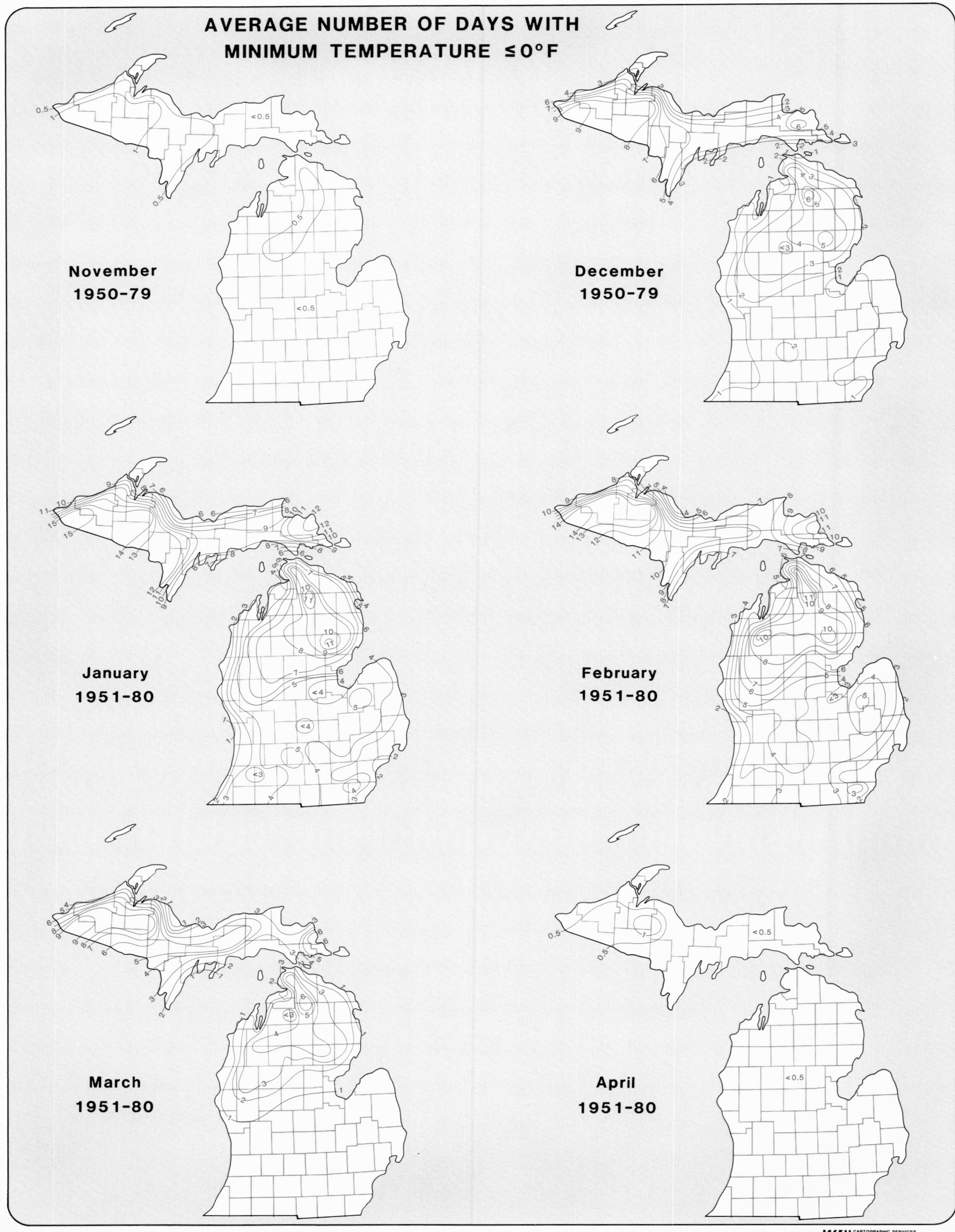


Source: MDA/Climatology Program

WMU CARTOGRAPHIC SERVICES DEPARTMENT OF GEOGRAPHY

# Extremes of Temperature

The maps of extreme maximum and extreme minimum temperatures differ in one very important aspect from the others in this atlas. These two maps do not cover a specific period of time, but instead are for the total period through 1987 for each station with 20 or more years of record. Thus, if a station was not operative during a particularly hot or cold year, its extremes may be more moderate than a nearby station with a longer term of record. Thus the map patterns may indicate values that are not directly comparable due to the variation of time period for individual stations.

The maps do show that proximity to the Great Lakes exerts a strong control over the extremes of temperatures recorded in Michigan. In some cases, this control may be more important than latitude. Most stations except those along the immediate shores of the lakes have experienced summer extremes of over 100°F. All stations except two in the Detroit metropolitan area have recorded winter extremes of less than −20°F. It is interesting to note that both the statewide extreme maximum (112°F, Mio, July 13, 1936) and minimum (−51°F, Vanderbilt, February 9, 1934) temperatures were recorded in the northern interior portion of the Lower Peninsula, within 40 miles and within three years of each other.

In addition to the highest temperature of record in Michigan of 112°F at Mio, Saginaw and Newaygo recorded 111°F, Kalamazoo and Hastings recorded 109°F, and eight other stations recorded 108°F on that same day. Mio, Bay City, Saginaw, and Newaygo are the only stations to record a temperature above 110°F. The lowest extreme high temperature is 93°F at Cross Village, and 94°F at Kincheloe AFB/Rudyard 4N.

In addition to the lowest temperature of record at Vanderbilt, a −51°F, Baldwin in the west central Lower Peninsula had −49°F. In the western interior portion of the Upper Peninsula, Bergland Dam has recorded −48°F and Watersmeet and Stambaugh −47°F. Seney, in the eastern Upper Peninsula, has also recorded −47°F. The highest extreme low temperature is −16°F at Grosse Pointe Farms.

A comparison of the two extreme temperature maps reveals a striking difference in patterns in the Lower Peninsula. In some areas the maximum temperature isotherms parallel the Lake Michigan shoreline much more than do the minimum temperature isotherms. This is probably due to the combined influence of local topography and stable light wind/calm conditions that accompany extremely cold conditions. Thus the lake influence is minimal and the influence of local relief is maximized under these conditions.

By contrast, the extremely high temperatures promote atmospheric instability and the generation of lake breezes. Thus, lake influence is enhanced and the effects of local relief are minimized under these conditions. In addition, some of the extreme minimum temperatures have been recorded with light offshore winds thus completely negating any potential lake influence.

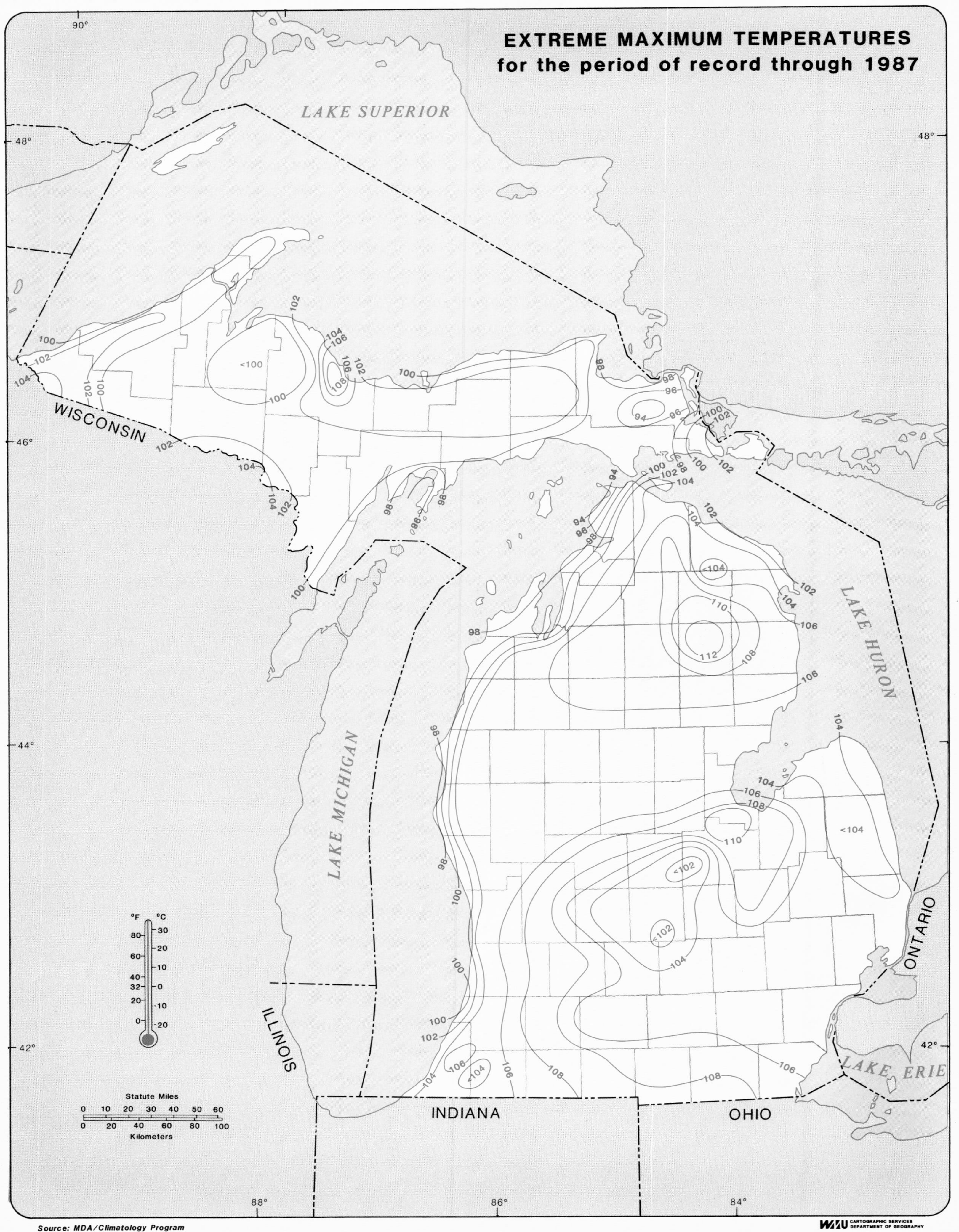
EXTREME MAXIMUM TEMPERATURES
for the period of record through 1987
LAKE SUPERIOR
LAKE MICHIGAN
LAKE HURON
LAKE ERIE
WISCONSIN
ILLINOIS
INDIANA
OHIO
ONTARIO
Statute Miles
Kilometers
°F
°C
Source: MDA/Climatology Program
WMU CARTOGRAPHIC SERVICES DEPARTMENT OF GEOGRAPHY

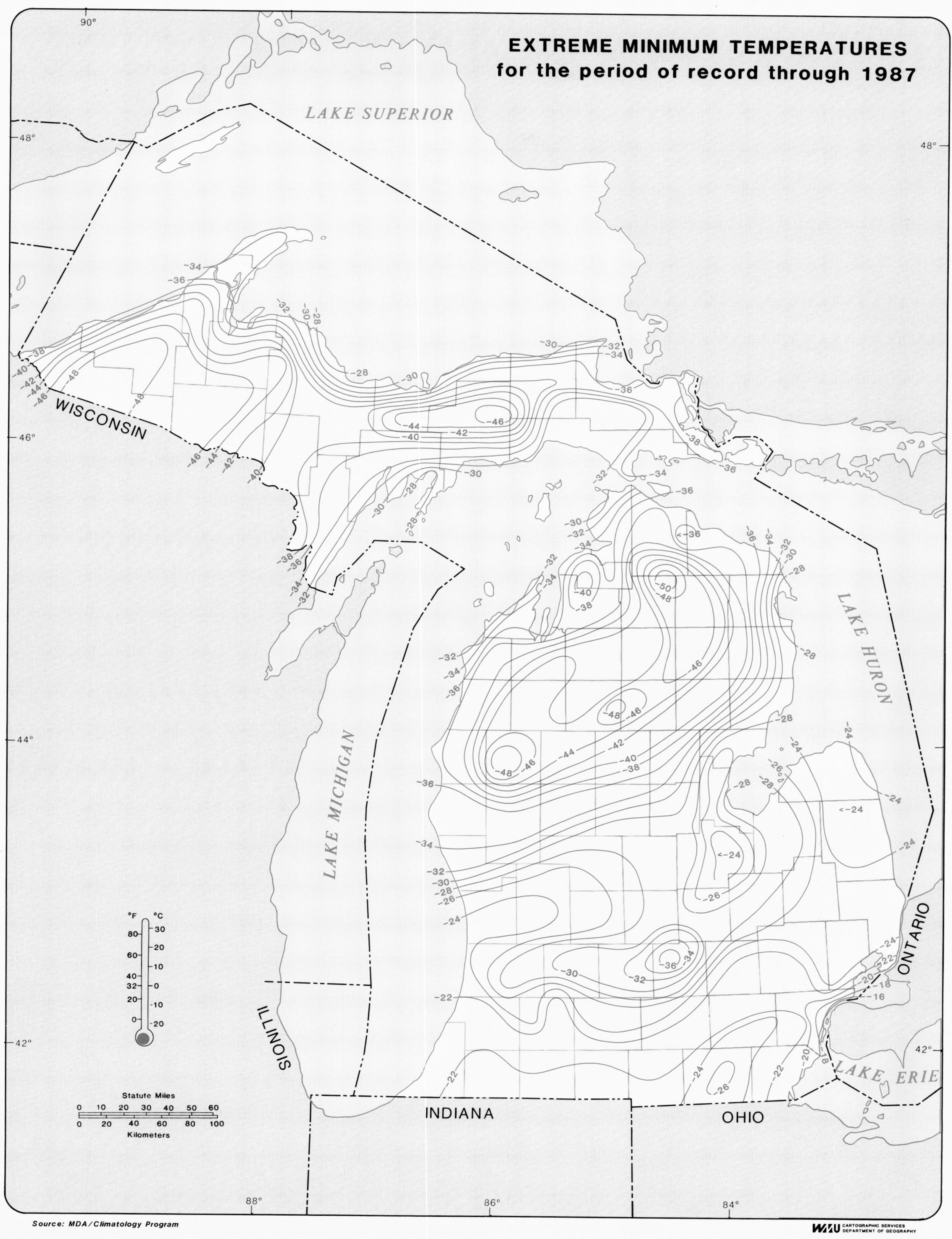


Source: MDA/Climatology Program

WMU CARTOGRAPHIC SERVICES DEPARTMENT OF GEOGRAPHY

# 32°F Threshold Dates/Seasons

The usual occurrence of freezing temperatures ending in the spring and beginning in the fall sets general limits to agricultural production in Michigan. Responding to lake proximity as well as to latitude and local topography, the growing season (number of days between last spring 32°F on or before July 31 and first fall 32°F temperatures after July 31) varies greatly within the state and from year to year. The average length of the growing season in Michigan ranges from less than 70 days in the interior of the northern Lower Peninsula to more than 170 days in the extreme southwestern and southeastern portions of the Lower Peninsula. The shortest growing season experienced during the study period* was 3 days at Vanderbilt, and the longest growing season was 225 days at Grand Rapids Kent County Airport. It is noteworthy that the shortest average growing season occurs in the northern part of the Lower Peninsula, not the Upper Peninsula. This is because the cold pole of daily minimum temperature shifts from the Upper Peninsula during the cold season to the northern Lower Peninsula during the warm season.

The average date of the last spring frost varies from June 20 at Vanderbilt to April 24 at Detroit City Airport, although from 1930–1979 the date of the last spring frost ranged from a earliest date of March 30 at Benton Harbor and Detroit City Airport to a latest date of July 31 at Vanderbilt.

The average date of the first fall frost in Michigan comes August 22 at Vanderbilt, but not until October 23 at Detroit City Airport. During 1930–1979 the dates have ranged from as early as August 3 at Vanderbilt and Grand Marais to a late November 29 at Grand Rapids.

*The user should note that a time period of 1930–1979 has been used in this and the following section.

# 28°F Threshold Dates/Seasons

The occurrences of 28°F temperatures in the spring and fall may be, for some crops, even more critical than 32°F temperatures. Crops such as tree fruit may be able to withstand freezing temperatures but succumb when temperatures fall to 28°F or below. The critical temperatures are highly dependent on the stage of growth, however.

The average number of days between last spring and first fall 28°F occurrences in the 1930–1979 period ranges from 101 days at Vanderbilt to 210 days at Detroit City Airport. The least number of days during 1930–1979 occurred at Watersmeet with 61 and the largest number of days at Detroit City Airport with 250.

The average date of the last spring 28°F occurrence ranges from a late June 3 at Vanderbilt to an early April 10 at Detroit City Airport. The latest and earliest dates that occurred in the 1930–1979 period are July 6 at Vanderbilt and March 9 at Detroit City Airport.

The average date of the first fall 28°F occurrence ranges from September 13 at Vanderbilt, to November 6 at Detroit City Airport. From 1930–1979, 28°F temperatures first occurred as early as August 18 at Watersmeet and as late as December 2 at Benton Harbor and Grand Rapids.

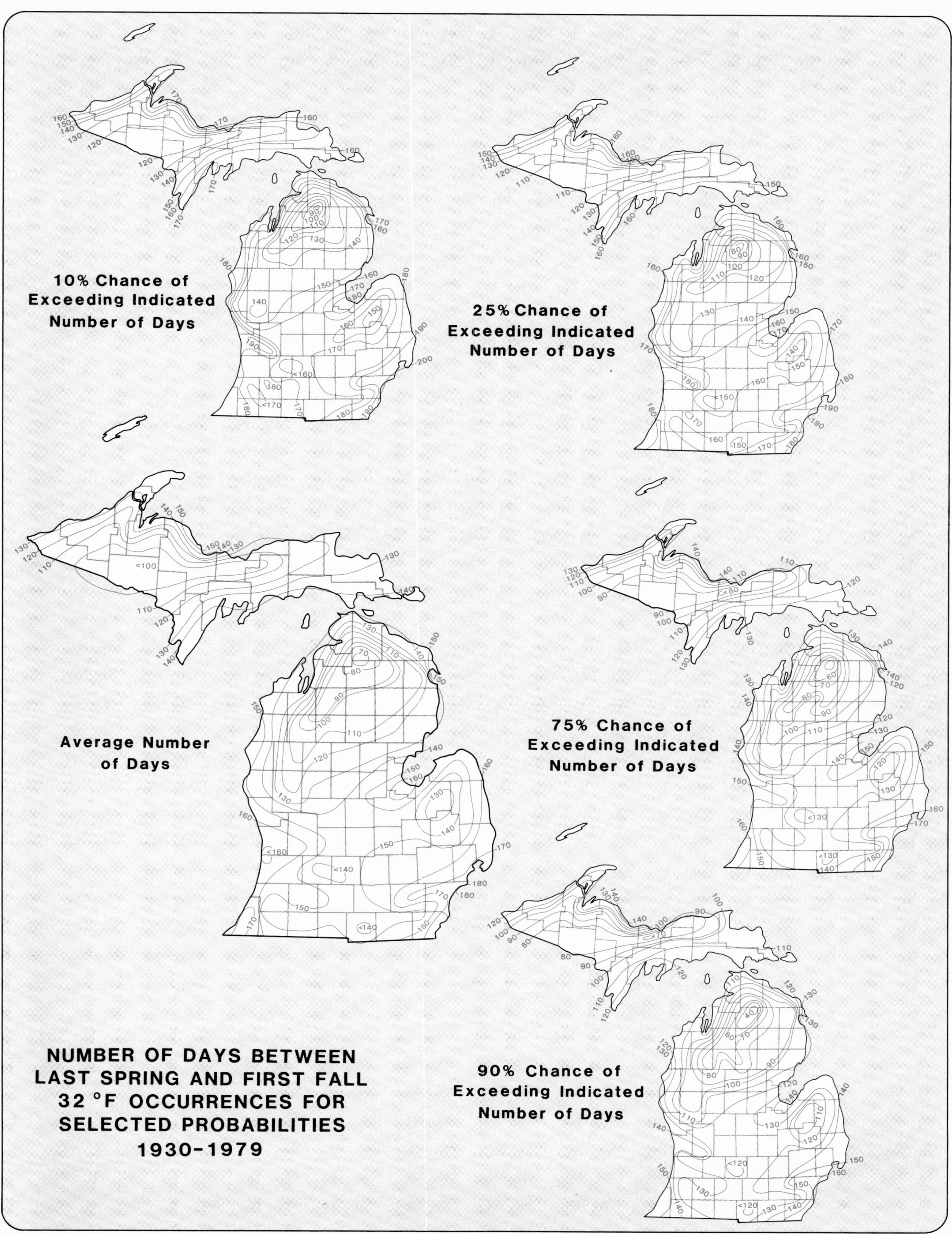


Source: MDA/Climatology Program

WMU CARTOGRAPHIC SERVICES DEPARTMENT OF GEOGRAPHY

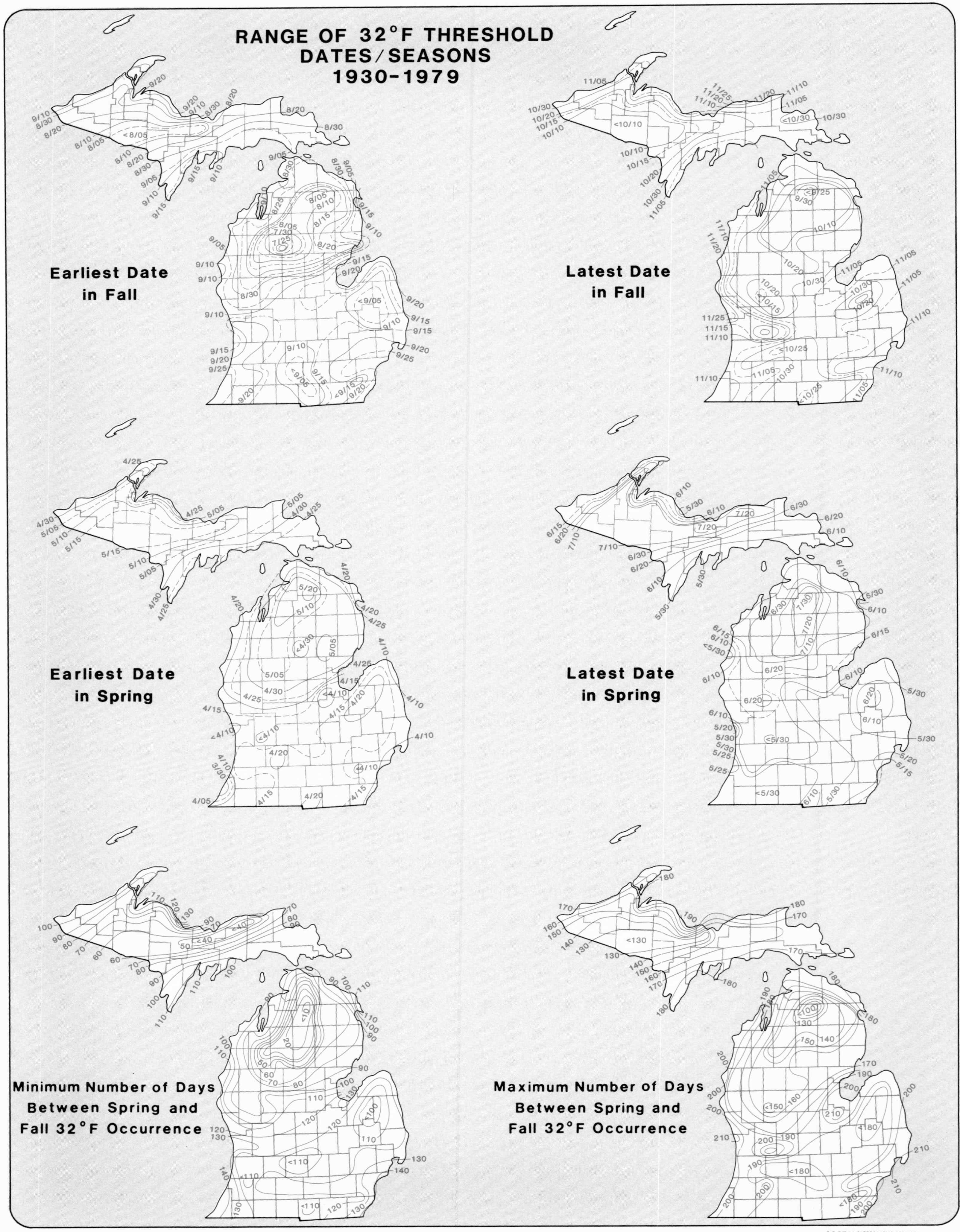


*Source: MDA/Climatology Program*

WMU CARTOGRAPHIC SERVICES DEPARTMENT OF GEOGRAPHY

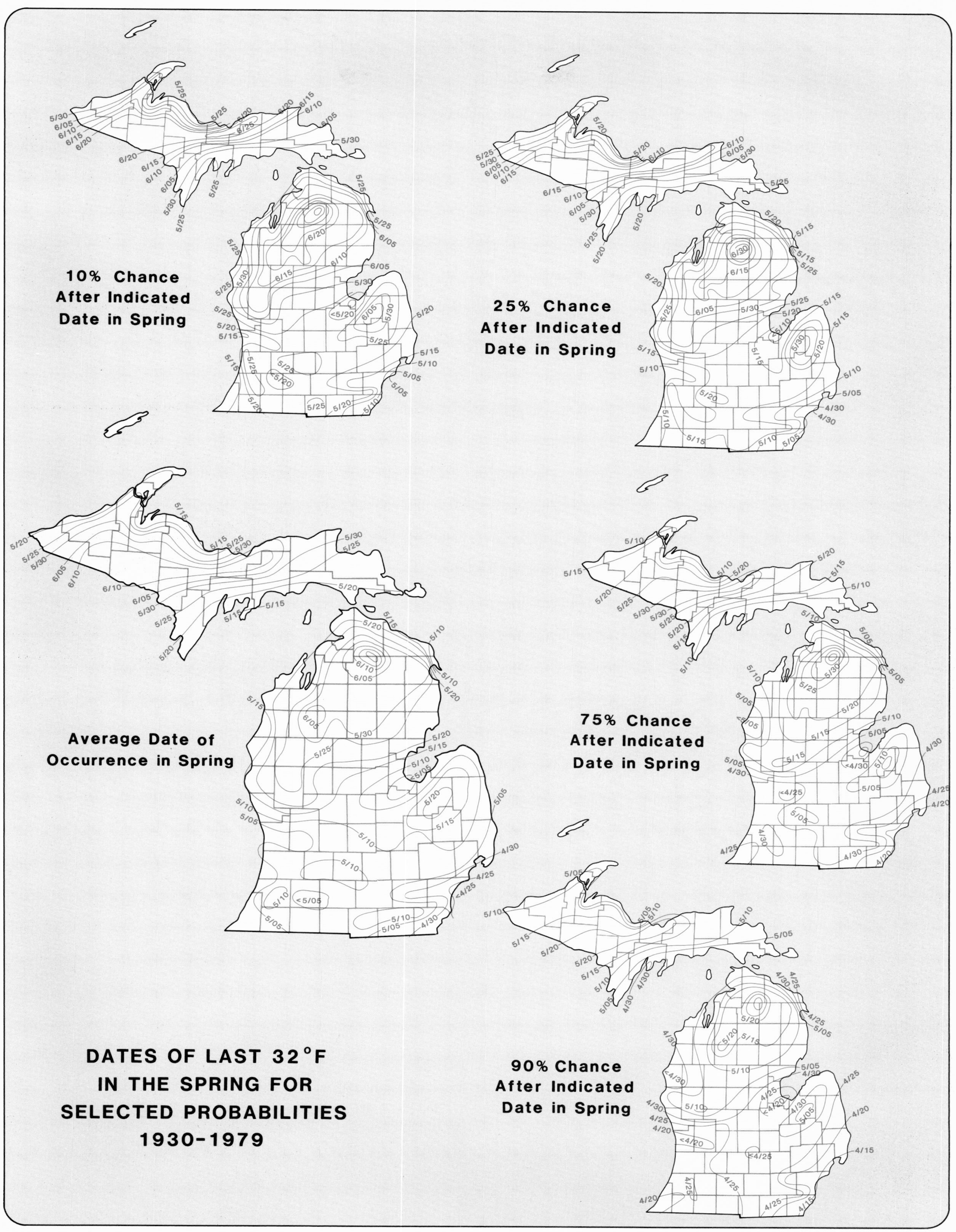


*Source: MDA/Climatology Program*

WMU CARTOGRAPHIC SERVICES DEPARTMENT OF GEOGRAPHY

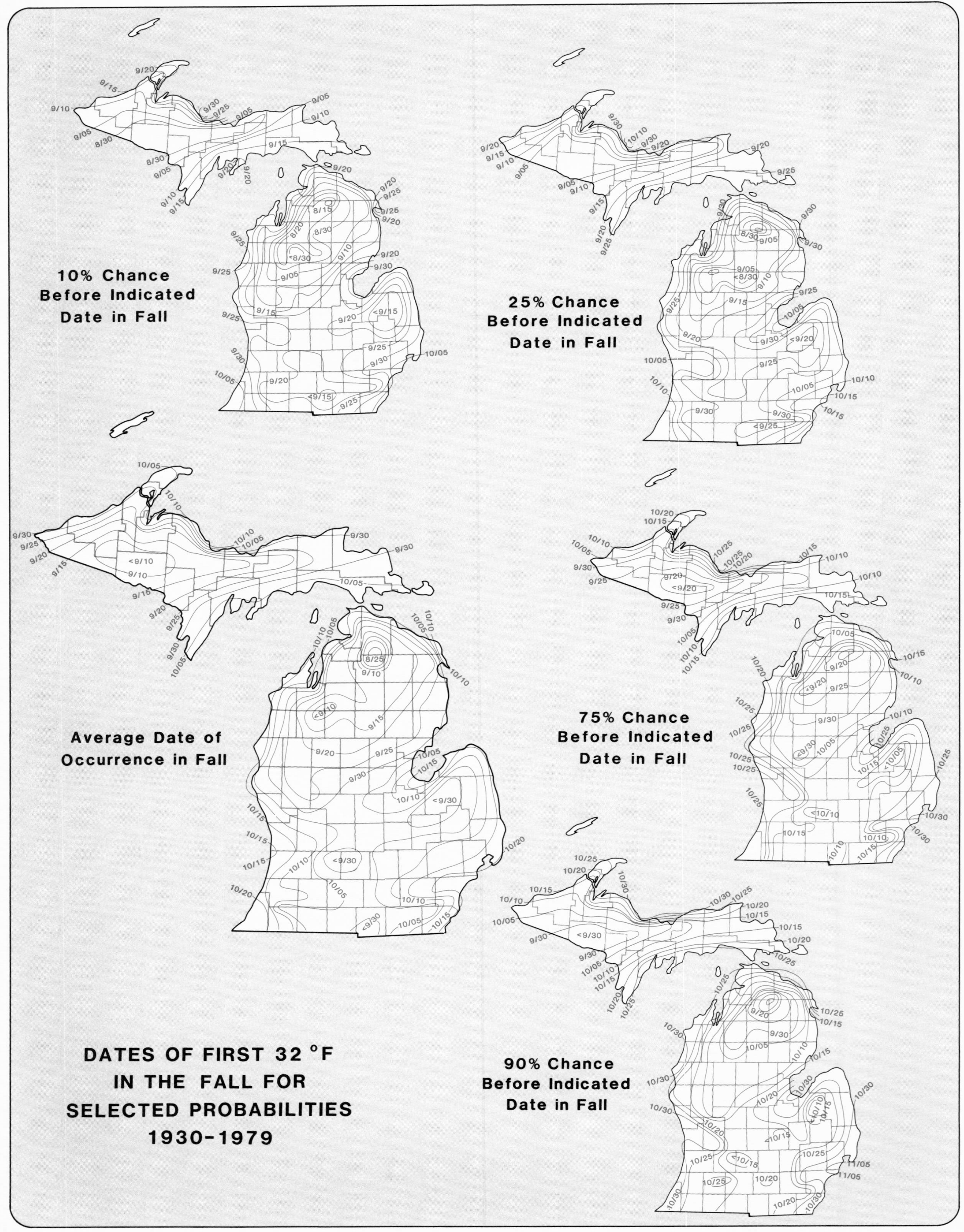


*Source: MDA/Climatology Program*

WMU CARTOGRAPHIC SERVICES DEPARTMENT OF GEOGRAPHY

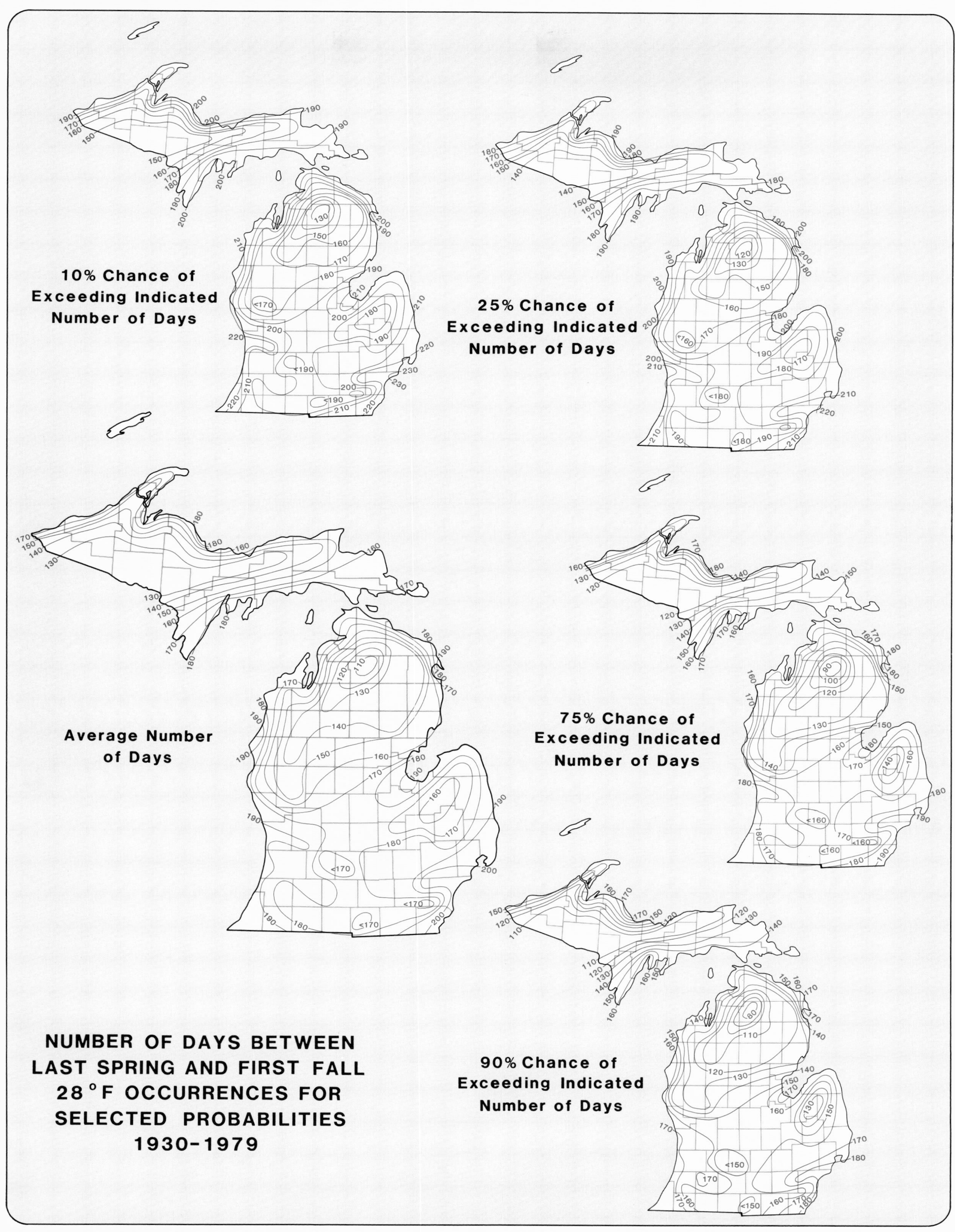


*Source: MDA/Climatology Program*

WMU CARTOGRAPHIC SERVICES DEPARTMENT OF GEOGRAPHY

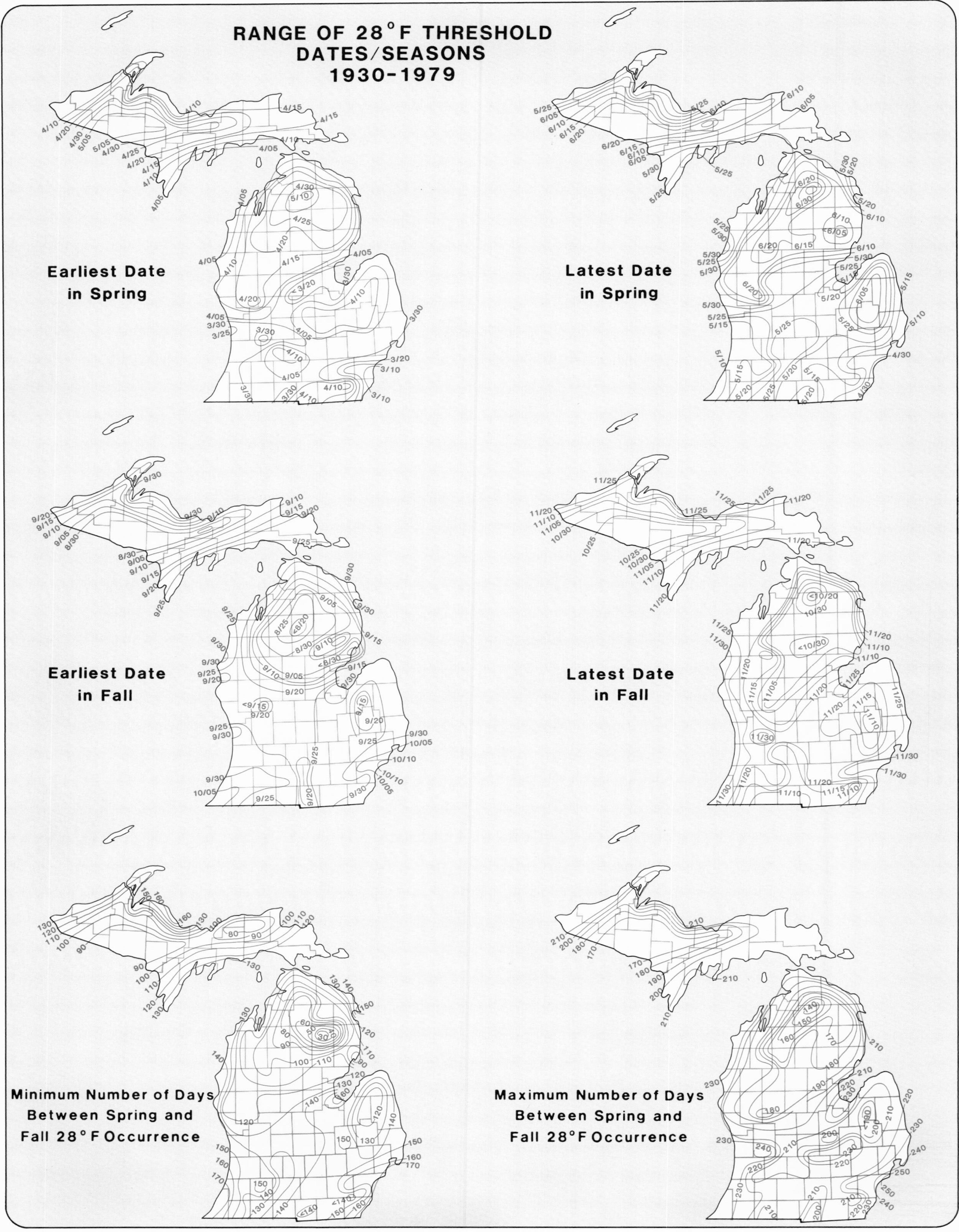


Source: MDA/Climatology Program

WMU CARTOGRAPHIC SERVICES DEPARTMENT OF GEOGRAPHY

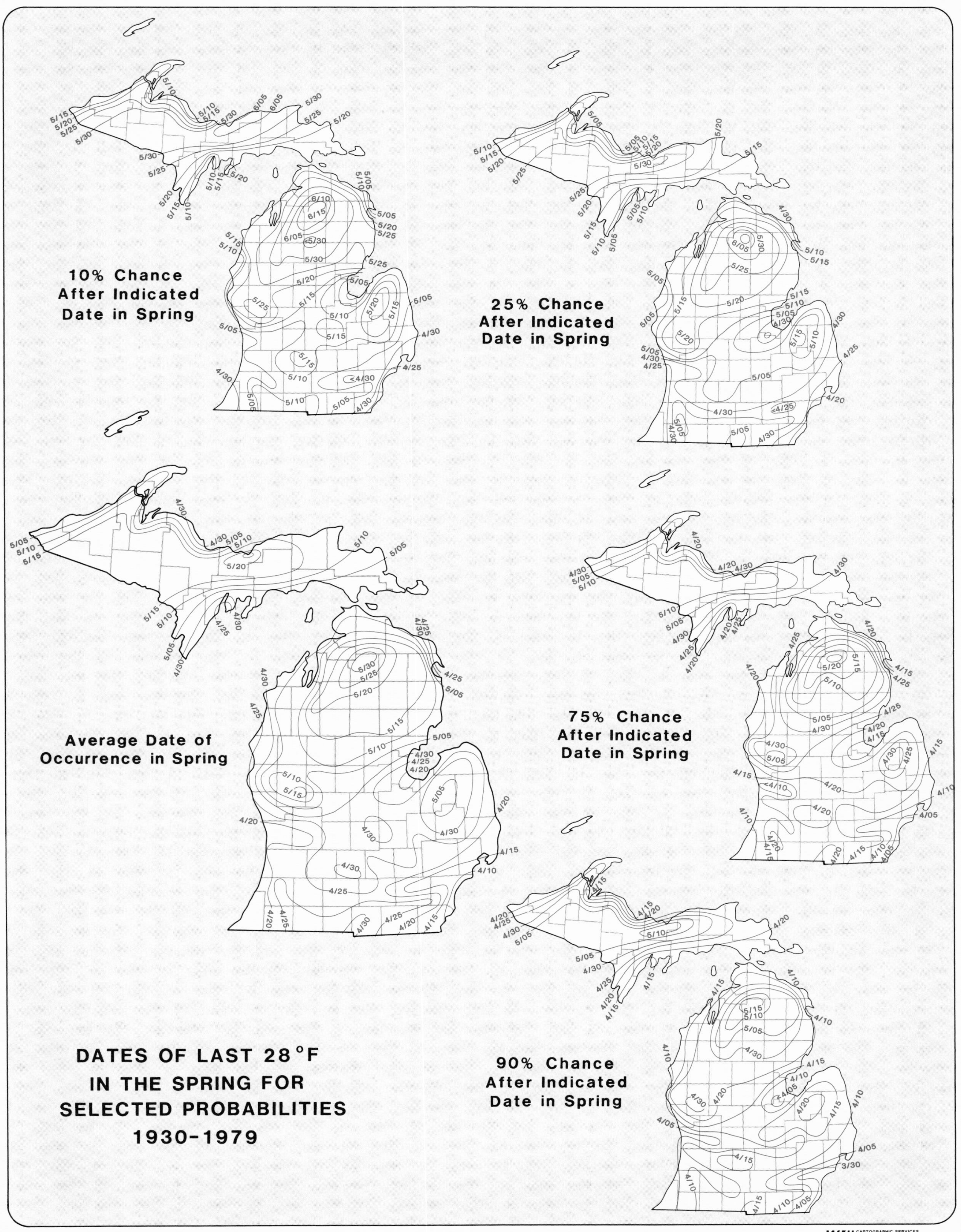


*Source: MDA/Climatology Program*

WMU CARTOGRAPHIC SERVICES DEPARTMENT OF GEOGRAPHY

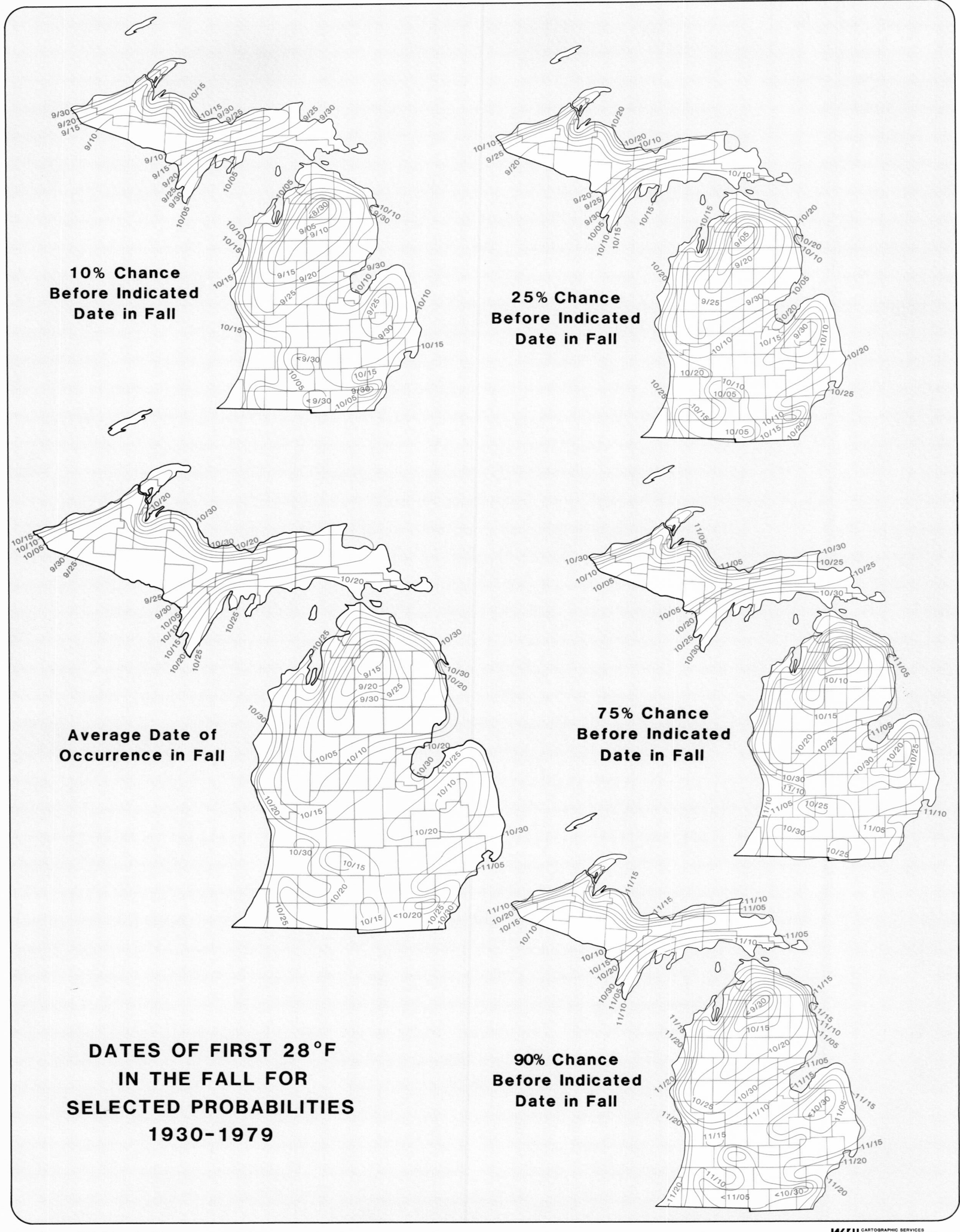


DATES OF FIRST 28°F
IN THE FALL FOR
SELECTED PROBABILITIES
1930-1979

Source: MDA/Climatology Program

WMU CARTOGRAPHIC SERVICES DEPARTMENT OF GEOGRAPHY

Fall color in Michigan.

# Heating Degree Days

## *Seasonal (July-June)*

The accumulation of "heating degree days" over a period of time provides a comparative estimate of the amount of energy required for domestic heating. A base temperature of 65°F has been found to be a threshold value, above which daytime heating is not needed. Energy consumption for heating has been found to increase directly with the departure of average daily temperatures below the base value of 65°F.

Each degree that the mean daily temperature drops below the base value of 65°F is called a "heating degree day". For example, at Station A, where the mean daily temperature is 60°F, the number of heating degree days is 5. If station B has as mean daily temperature of 55°F, the number of heating degree days is 10. Station B would thus require twice as much energy for heating to human comfort as Station A.

Averages of accrued heating degree days for the heating season (July-June), or for individual months may be computed by summing negative departures of the average daily means from the base value of 65°F. Seasonal or monthly values of heating degree days may then be compared from place to place to derive estimates of energy consumption. For example, seasonal average accumulations of heating degree days in the conterminous United States range from a high of 10,000 in northern Minnesota to a low of 100 in the Florida keys.

In general, 50th percentile seasonal heating degree accumulations in Michigan range from the upper 6000s in the southern Lower Peninsula to the lower 9000s in the western Upper Peninsula. This points to a 35% difference in energy consumption between the two areas. The absolute range is from a maximum of 9625 at Champion in the Upper Peninsula to a minimum of 6202 at Detroit City Airport. Thus, the seasonal energy requirement for heating would be 55% larger at Champion.

## *Monthly*

All stations in Michigan accumulate heating degree days during all months of the year with the exception of Detroit City Airport in July. This indicates some need within the state for indoor heating during all months of the year.

The heating season begins in July, although values are small, ranging from 0 at Detroit City Airport to 137 at Whitefish Point. Monthly heating degree days increase during late summer and fall to reach maximum values in January, normally the coldest month. In January, the 50th percentile values range from 1228 at Grosse Pointe Farms to 1700 at Bergland Dam, indicating 38% more energy consumption at the Upper Peninsula station.

Monthly 50th percentile values then decrease through late winter and spring. In June, at the end of the heating season, the values range from 298 at Whitefish Point to 32 at Monroe.

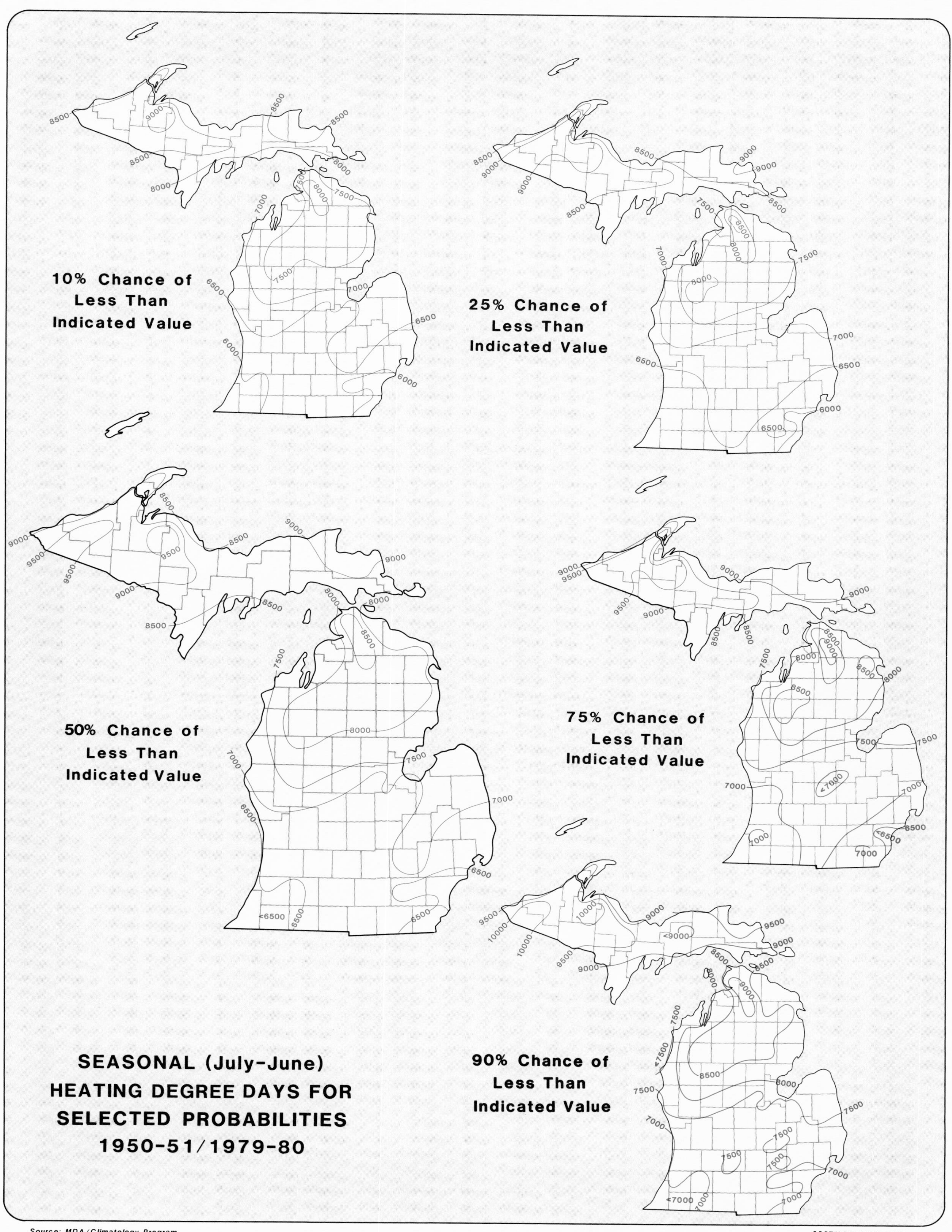


*Source: MDA/Climatology Program*

WMU CARTOGRAPHIC SERVICES DEPARTMENT OF GEOGRAPHY

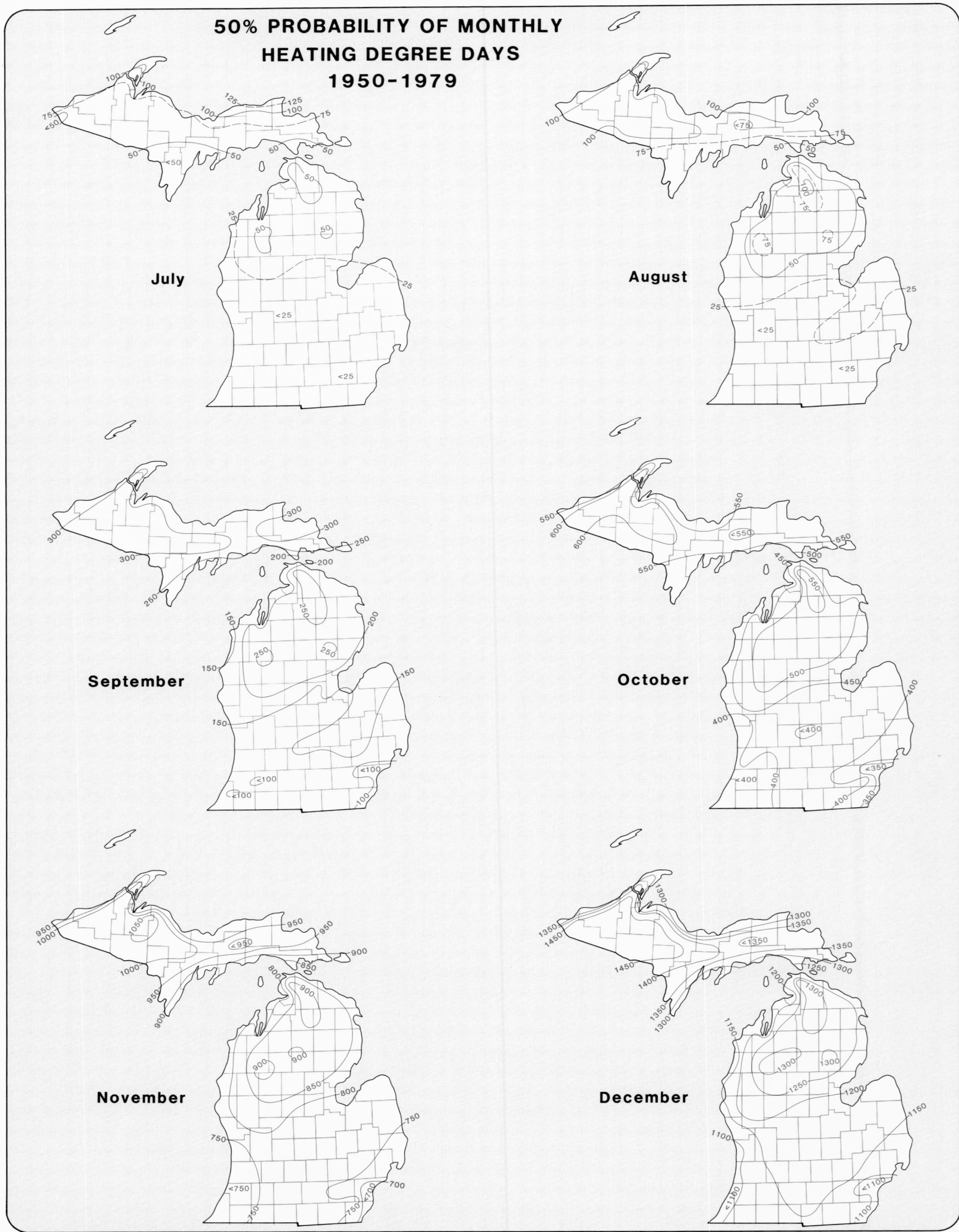


*Source: MDA/Climatology Program*

WMU CARTOGRAPHIC SERVICES DEPARTMENT OF GEOGRAPHY

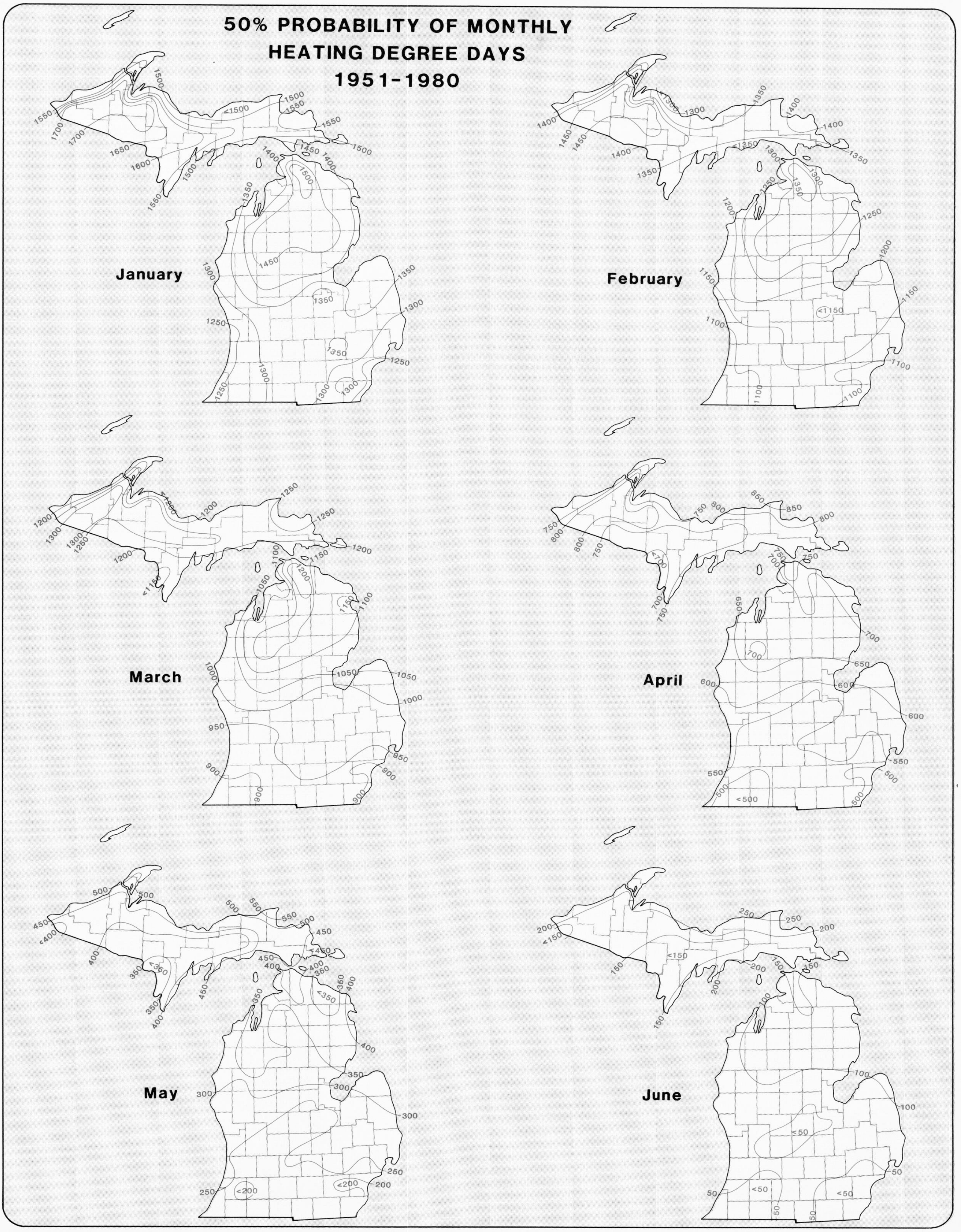


*Source: MDA/Climatology Program*

WMU CARTOGRAPHIC SERVICES
DEPARTMENT OF GEOGRAPHY

# Cooling Degree Days

## *Seasonal* (*May-October*)

Michigan summers are warm enough so that on some days there is a need to cool the indoor climate for comfort. The accumulation of "cooling degree days" over periods of time provides comparative estimates of energy expenditures for air conditioning, just as heating degree days give estimates of energy needs for heating. However, due to larger sun angles and more solar radiation, the predictive nature of the "cooling degree day" is not as good as its heating counterpart.

The "cooling degree day" is computed in the same manner as the heating degree day, except the departure of mean daily temperature is above rather than below the base value of 65°F. Thus a day with a mean temperature of 70°F has an accumulation of 5 cooling degree days.

The 50th percentile values for the cooling season in Michigan (May-October) show that all stations in the state normally have some requirements for indoor cooling. The need may not be economically met by mechanical means, however, but by simply allowing for proper ventilation and shading. Cooling degree day seasonal accumulations range from over 800 in the extreme southeast part of the Lower Peninsula to less than 200 in portions of the Upper Peninsula. The largest value, 843, was at Detroit City Airport, while the smallest value, 72, was at Whitefish Point.

The relatively small seasonal values of cooling degree days as compared to heating degree days is not surprising considering Michigan's northerly location. For example, at Detroit City Airport the energy required for heating annually would be over seven times that required for cooling. An average cooling degree accumulation in south Florida of 4000, however, would place cooling energy requirements in that area at nearly forty times the heating energy requirements.

## *Monthly*

Most stations in the southern Lower Peninsula accumulate a small number of cooling degree days in April. The 50th percentile values are generally 5 or less so are not significant on the seasonal scale. These amounts are not presented in map form but are included in the seasonal totals.

Significant cooling degree days normally begin to accumulate during May. However, a number of stations in the eastern portion of the Upper Peninsula did not experience cooling degree days in May for half of the years during the 1951–1980 period. All stations have accumulated cooling degree days in May in one or more years.

All stations can expect some cooling degree days in June, and maximum monthly values occur in July, the warmest month, at most stations, although the maximum at Whitefish Point occurs in August. By September, cooling degree days decrease substantially, although all stations have some. During October, which is the end of the cooling season, the accumulation of cooling degree days is highly variable just as the spring. All stations did accumulate cooling degree days in one or more years during the 1951–1980 period.

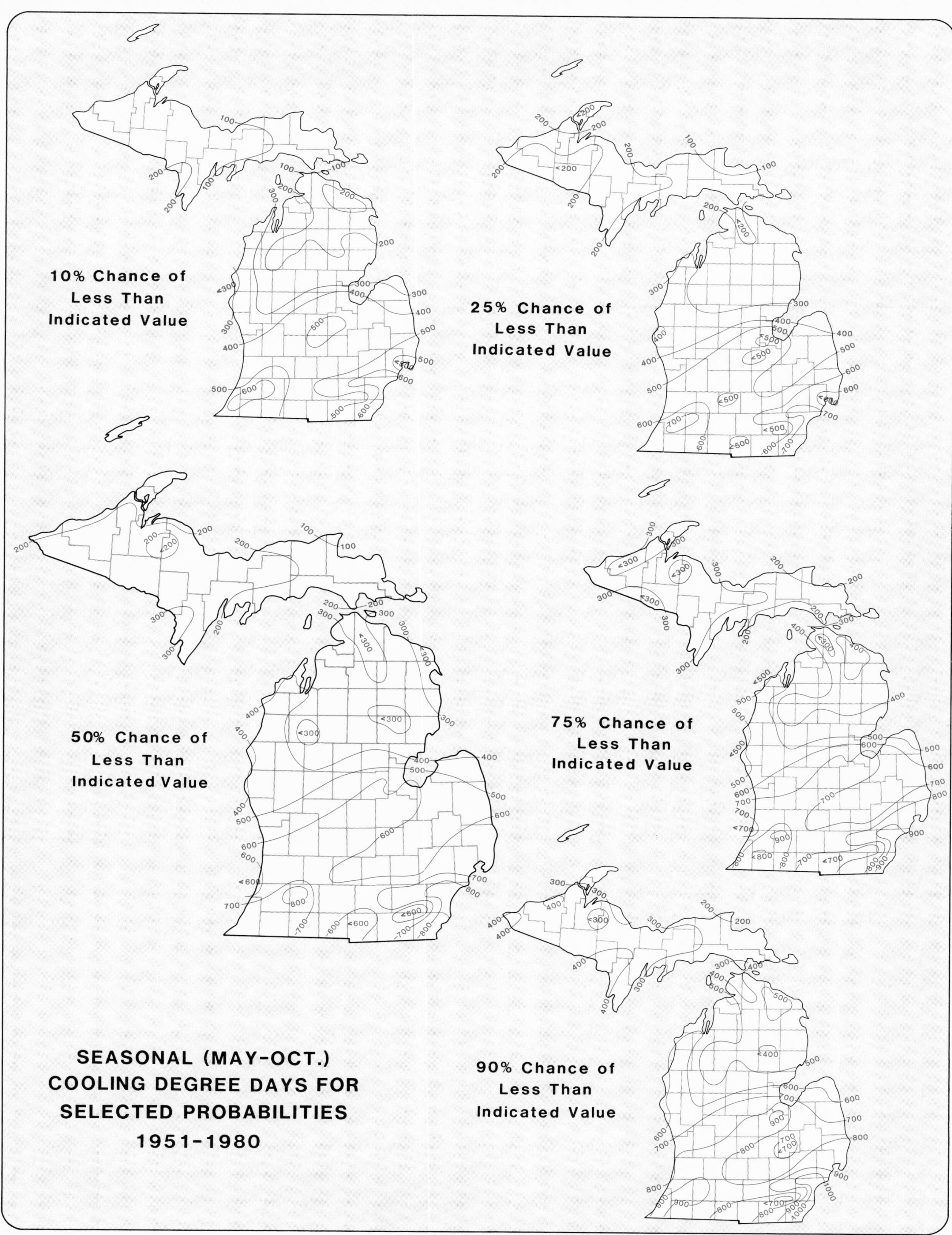


Source: MDA/Climatology Program

WMU Cartographic Services Department of Geography

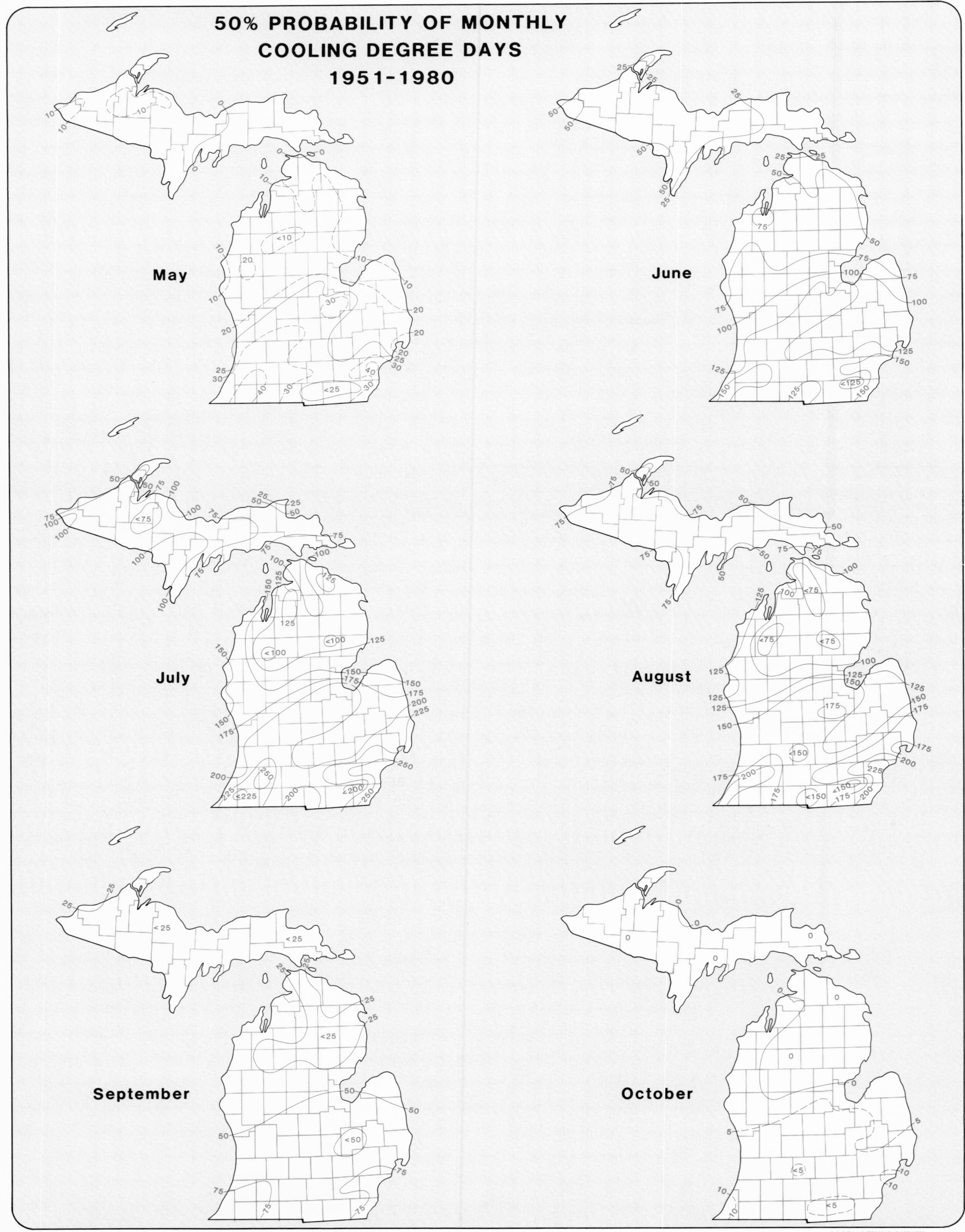


Source: MDA/Climatology Program

WMU CARTOGRAPHIC SERVICES DEPARTMENT OF GEOGRAPHY

Lake Superior, Whitefish Bay.

# Growing Degree Days

A "growing degree day" is an indicator of the amount of energy available for biological activity, i.e., plant growth, pest development, diseases, etc. Some examples of their use are:

1) Hybrid corn is rated as to the number of growing degree days to a base of 50°F that are required for maturity.
2) The maturity of various fruits can be forecast by knowing the growing degree days to selected bases.
3) Insect pest development can be predicted for optimum control procedures by the number of growing degree days to a certain base for that insect.
4) The scheduling of the planting of various vegetables, such as green beans, can be made by knowing the number of growing degree days needed for maturity and what will likely be the number available during various times of the season. Thus a processing plant can remain open longer and have the freshest possible vegetables to process.

These are but a few of the examples for which growing degree days are used.

The growing degree day is analogous to the cooling degree day in that it is the departure of the average daily temperature above a selected base temperature. The simplest form of the growing degree day, the one used here, uses the average daily temperature and then subtracts the selected base temperature. Thus for a day with an average temperature of 65°F the resulting growing degree days to a base of 50°F would be (65 − 50) = 15. These data are accumulated on a daily basis for all positive values for the duration of the growing season.

The base temperature used depends on the specific application. The ones chosen for presentation here, 40°F and 50°F, represent a range of values by which the user can see general patterns and make a decision if further investigation is warranted.

### *Seasonal (March-October) Growing Degree Days —Base 50°F*

The growing degree day to base 50°F season usually starts in March in the southern Lower Peninsula and progresses northward through April, peaks in July, and then decreases through October. Most spring and summer crops have been harvested by that time and the annual ornamental plants have been killed by the first freezes in the fall.

The 50th percentile of seasonal values during the 1951–1980 period ranged from lows of 1207 at Whitefish Point and 1441 at both Grand Marais and Sault Ste. Marie to highs of 3091 and 3077 at Dearborn and Kalamazoo, respectively.

### *Monthly*

The base 50°F growing season in Michigan starts in March. However, April is the first full month in the growing season statewide. During the remainder of the season through October, all stations accumulate growing degree days. In May the totals range from 31 at Whitefish Point to 331 at Kalamazoo. In June the range is from 166 at Whitefish Point to 583 at Monroe. In July, the warmest month of the year, the values range from 348 at Whitefish Point to 740 at Detroit City Airport. During August, the beginning of the summer cool-down, values range from 383 at Whitefish Point to 689 at Dearborn. Whitefish Point is the only station that has more growing degree days in August than in July, i.e., 383 vs. 348 during the 30-year period. During September the values range from lows of 172 at Houghton and 176 at Sault Ste. Marie to highs of 459 and 455 at Grosse Pointe Farms and Monroe, respectively. October is the ending month for summertime vegetative activity. The growing degree days range from lows of 41 and 43 at Sault Ste. Marie and Whitefish Point, respectively, to 176 and 172 at Grosse Pointe Farms and Detroit City Airport.

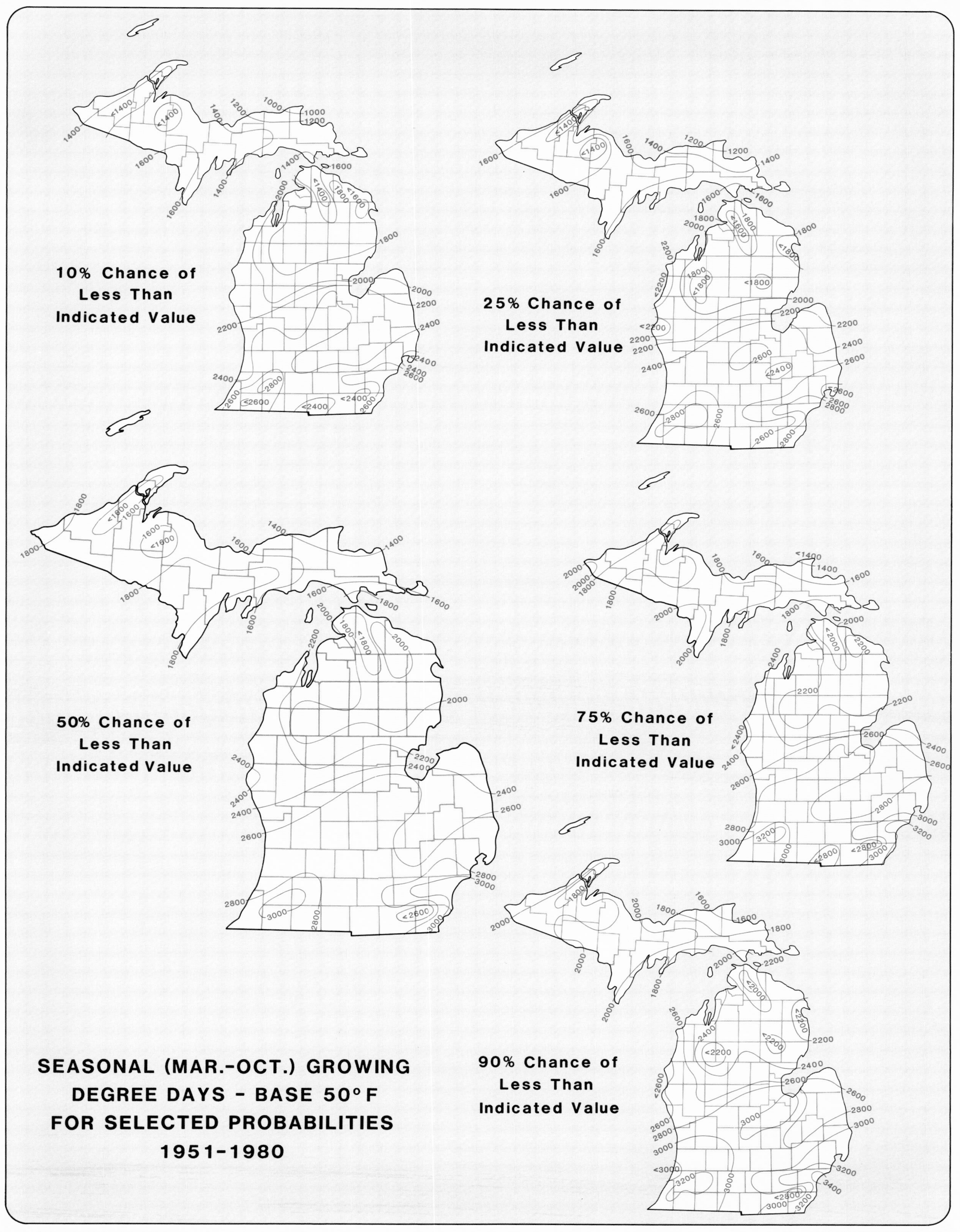


SEASONAL (MAR.-OCT.) GROWING DEGREE DAYS - BASE 50° F FOR SELECTED PROBABILITIES 1951-1980

Source: MDA/Climatology Program

WMU CARTOGRAPHIC SERVICES DEPARTMENT OF GEOGRAPHY

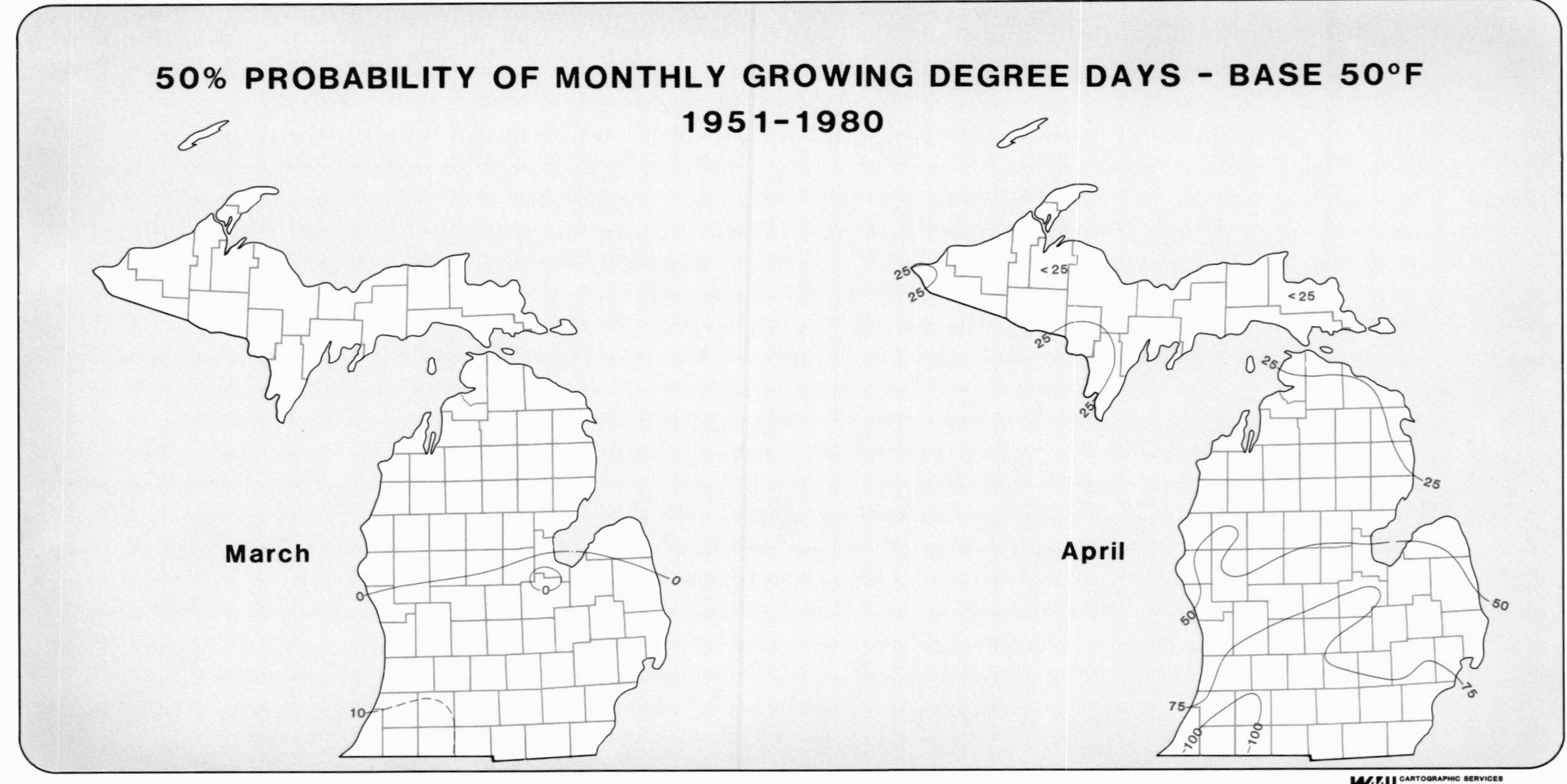
50% PROBABILITY OF MONTHLY GROWING DEGREE DAYS - BASE 50°F
1951-1980
March
0
0
0
10
April
25
25
< 25
< 25
25
25
25
25
50
50
75
75
100
100
Source: MDA/Climatology Program
WMU CARTOGRAPHIC SERVICES DEPARTMENT OF GEOGRAPHY

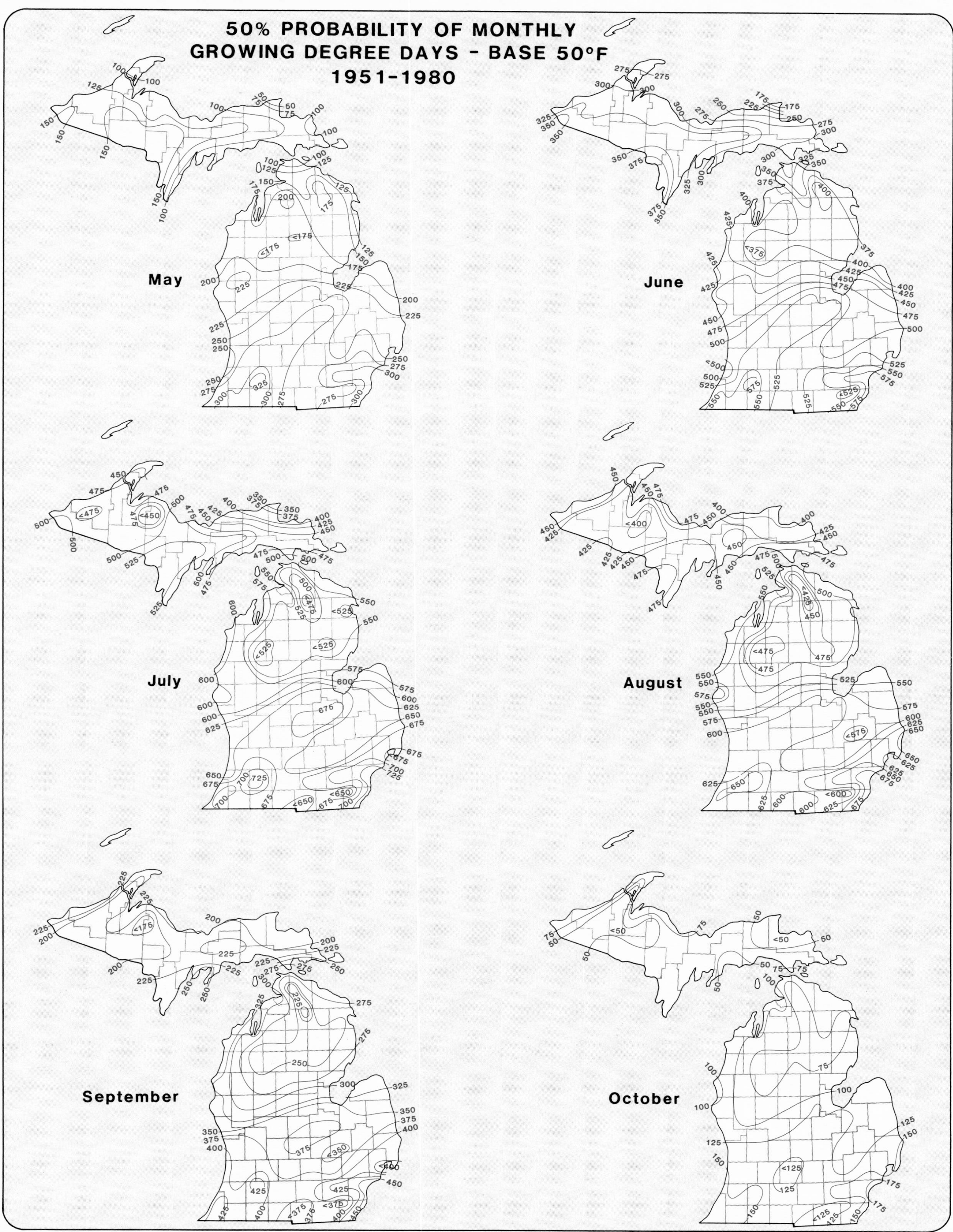


*Source: MDA/Climatology Program*

WMU CARTOGRAPHIC SERVICES DEPARTMENT OF GEOGRAPHY

## *Seasonal (March-October) Growing Degree Days —Base 40°F*

The growing degree day to base 40°F season starts in March throughout the state with no station failing to record growing degree days in at least one year in the 1951–1980 period. The season peaks in July and then decreases through October.

The 50th percentiles of seasonal values during the 1951–1980 period ranged from a low of 2786 at Whitefish Point to a high of 5100 at Dearborn.

### *Monthly*

April is the first full month of the growing season statewide. All stations with the exception of Whitefish Point recorded growing degree days on every year during the 30-year period. The 50th percentile values range from 25 at Whitefish Point to 295 at Kalamazoo. June values range from 452 at Whitefish Point to 887 at Monroe.

In July, the warmest month of the year, growing degree days range from 660 at Whitefish Point to 1050 at Detroit City Airport. During August, the beginning of the summer cool-down, values range from 695 at Whitefish Point to 1001 at Detroit City Airport. September values are largest at Grosse Pointe Farms with 761 and smallest at Houghton with 434.

October is the last month for summertime vegetative activity. The low values are at Champion (185) and the largest values are at Grosse Pointe Farms (435).

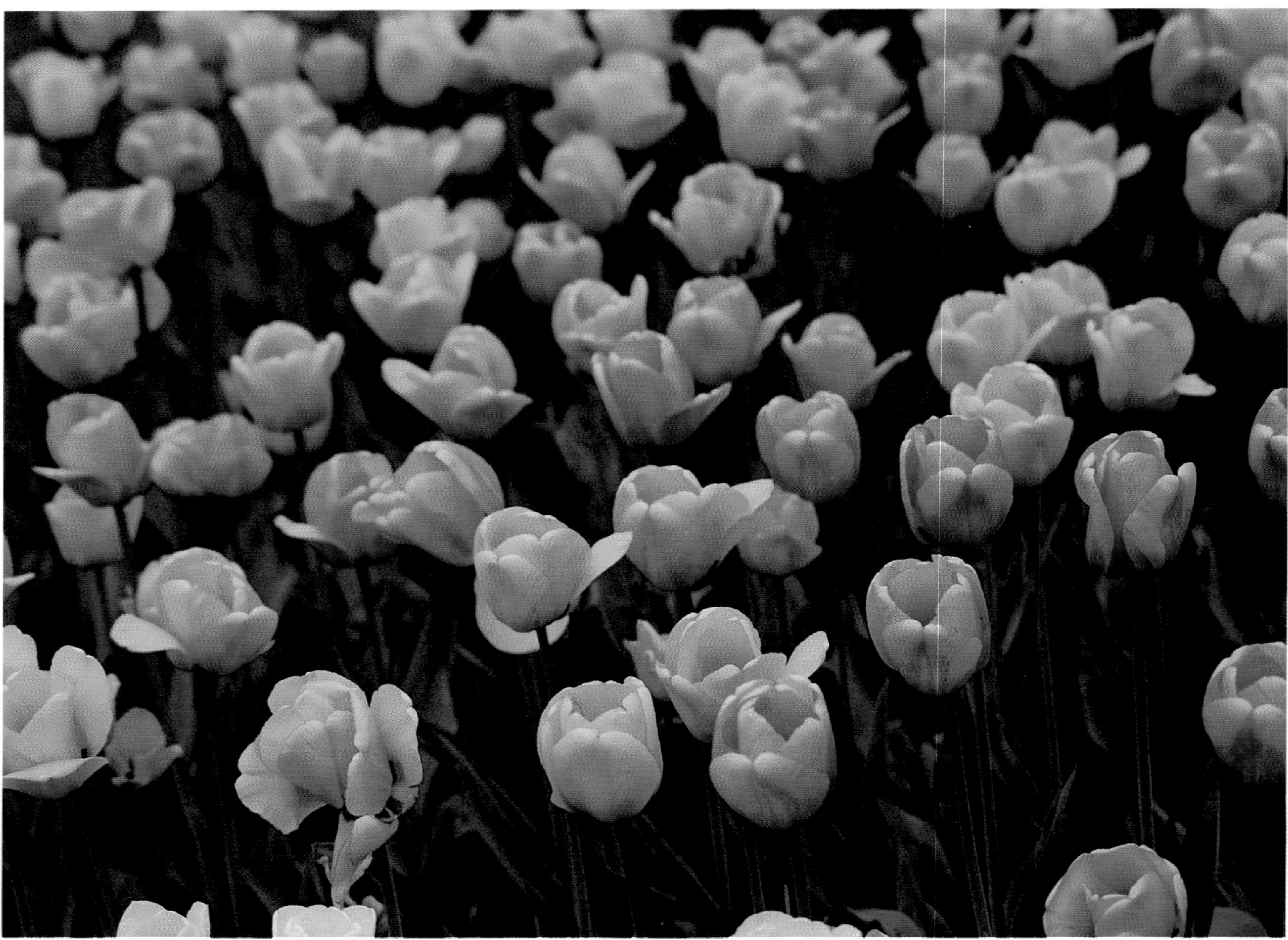

Tuliptime in Holland, Michigan.

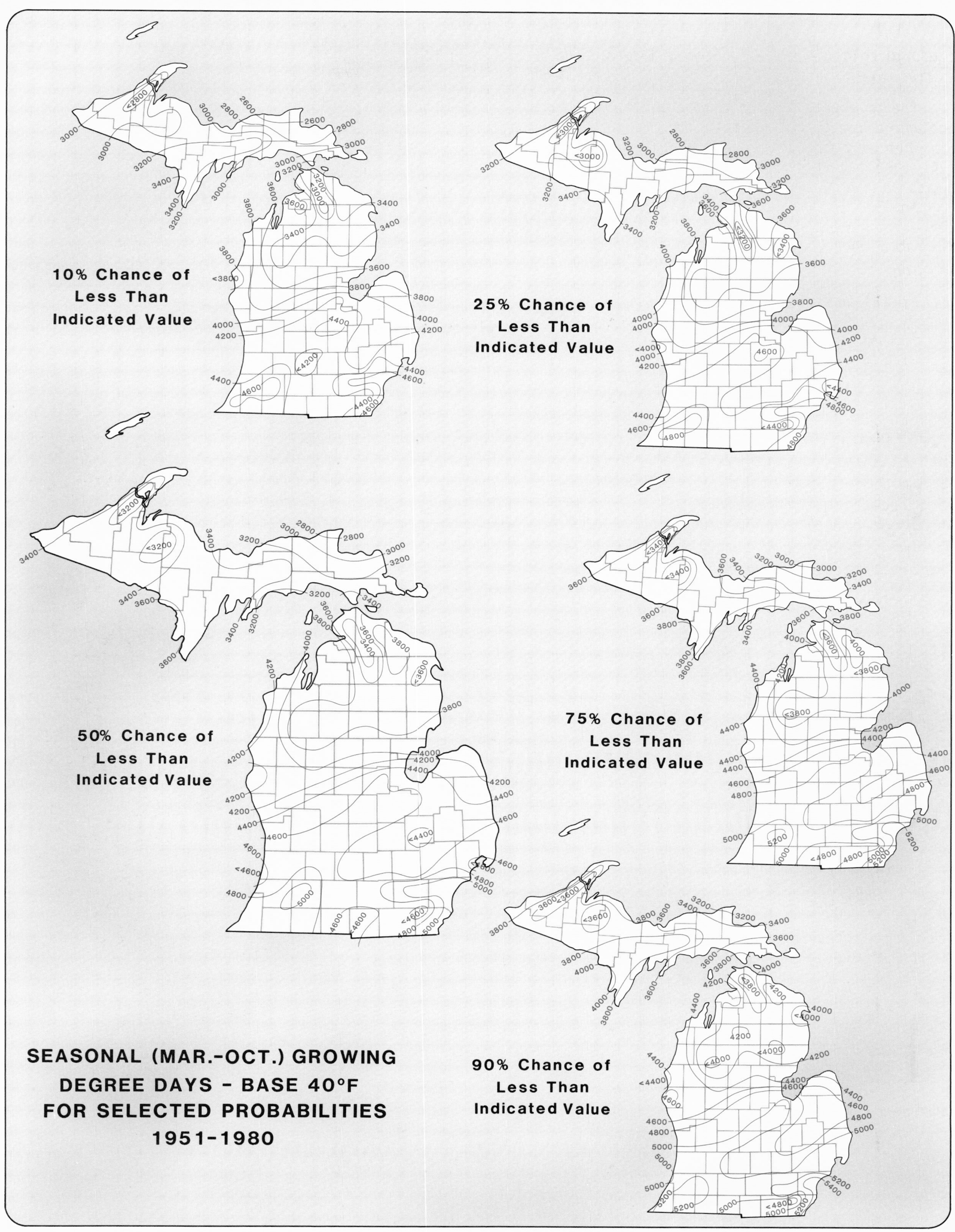


Source: MDA/Climatology Program

WMU CARTOGRAPHIC SERVICES DEPARTMENT OF GEOGRAPHY

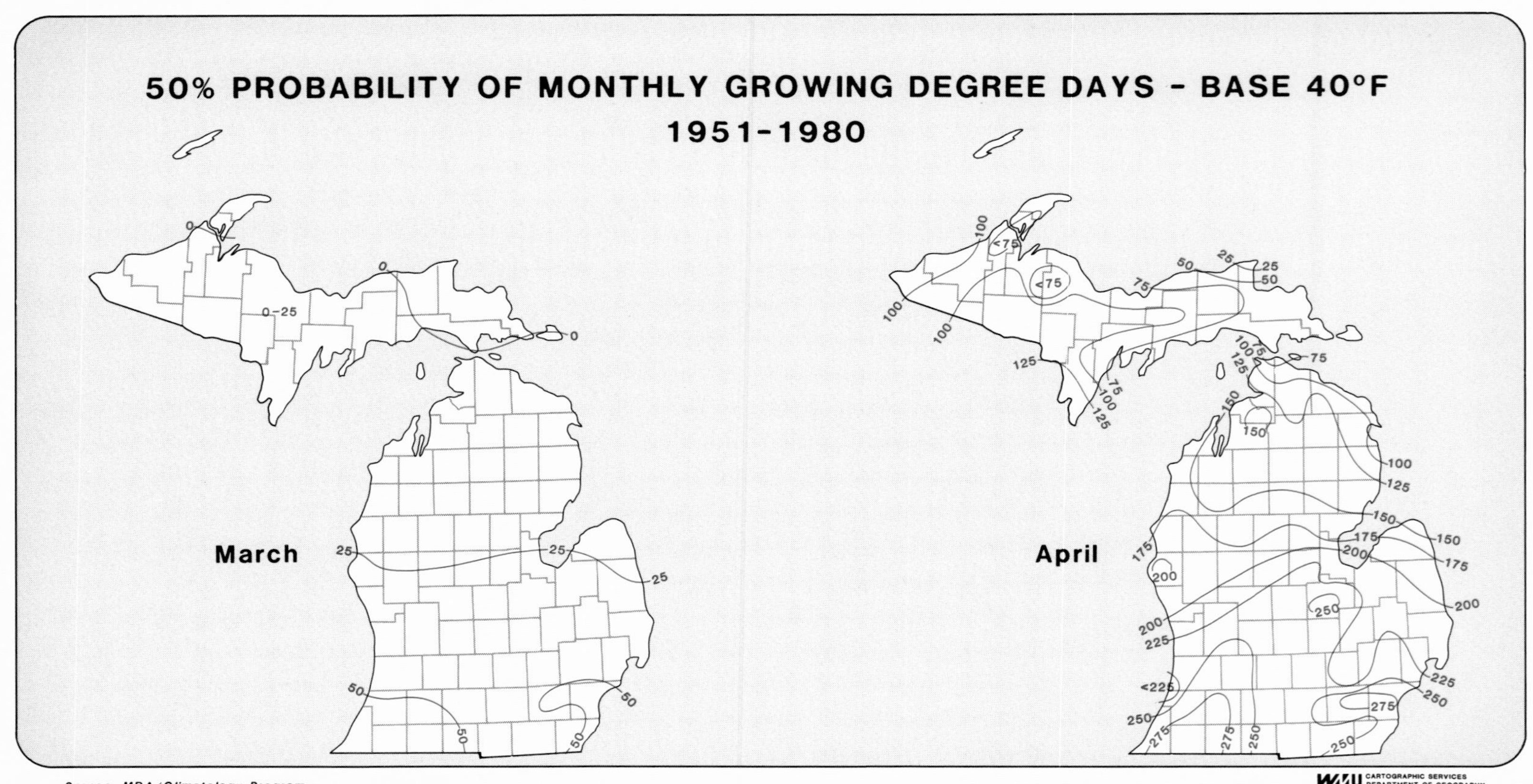
50% PROBABILITY OF MONTHLY GROWING DEGREE DAYS - BASE 40°F
1951-1980
March
April
Source: MDA/Climatology Program
WMU CARTOGRAPHIC SERVICES DEPARTMENT OF GEOGRAPHY

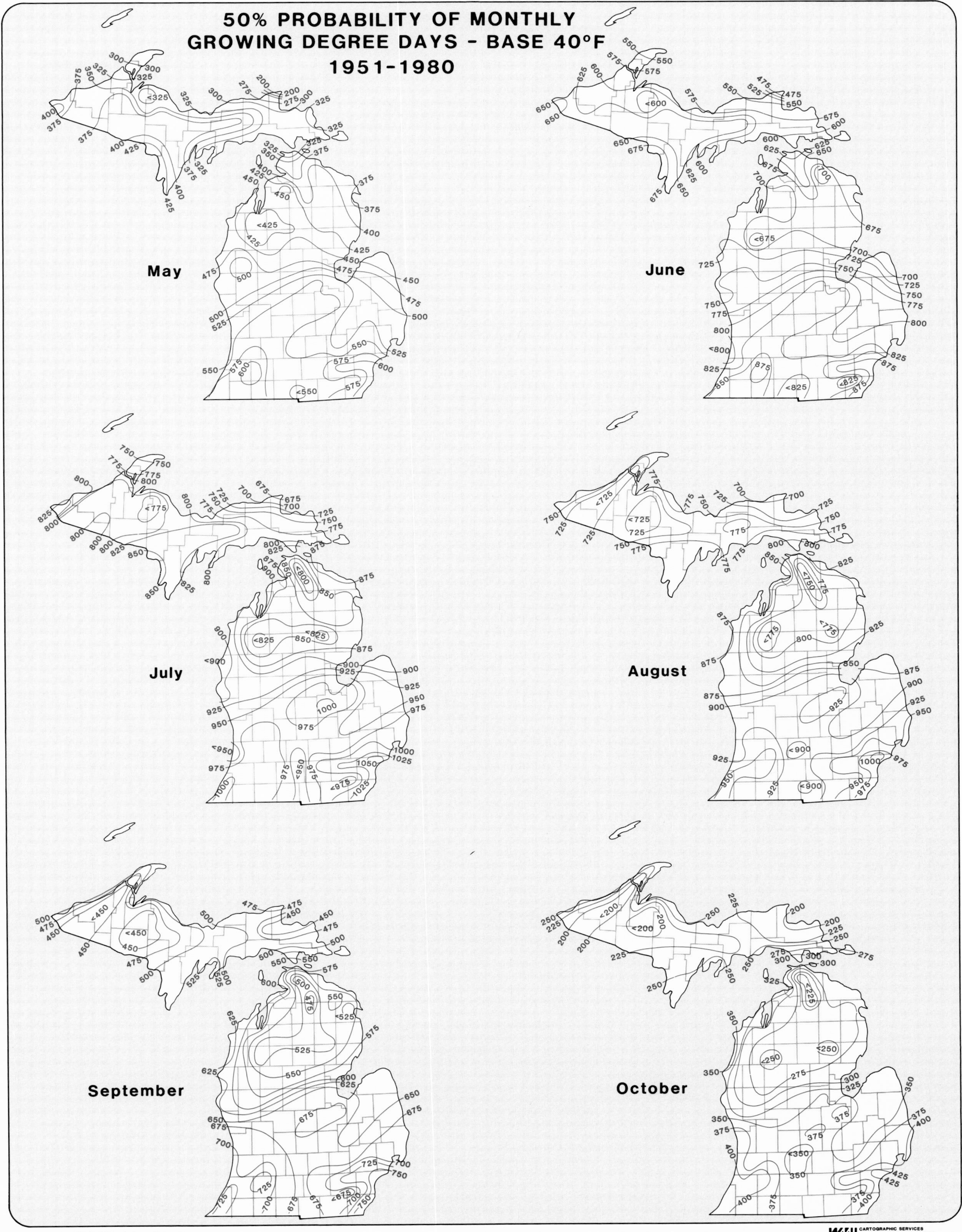


Source: MDA/Climatology Program

WMU CARTOGRAPHIC SERVICES DEPARTMENT OF GEOGRAPHY

# Average and Extreme Daily Temperatures

The seasonal progression of temperature in Michigan moves from a summer peak, occurring some weeks after the summer solstice, to a winter minimum, lagging after the winter solstice. However, long-period averages of daily temperatures show that the progression is not smooth. Warm spells and cold spells can be identified with specific dates and periods. Some spells, such as the "January Thaw," appear to occur over wide areas and have made a large imprint on weather folklore.

Daily averages of means, maxima, and minima, along with daily extremes, are shown graphically on the following pages for six Michigan stations. The data were archived and averages tabulated by the Michigan Department of Agriculture/Climatology Program. The time period of the data is variable.

The graphs clearly demonstrate the annual cycle of temperature. The coldest period is at the end of January and the beginning of February while the warmest period is in mid-July. The average daily range of temperatures (daily maximum minus daily minimum) is larger in summer than in winter.

The reader is advised *not* to try to interpret each oscillation in the lines as an indicator of something unique for that date. However, the frequency of oscillations does help to demonstrate the highly variable nature of climate in Michigan.

# AVERAGE DAILY TEMPERATURES

## Bay City (1887, 1896-1978)

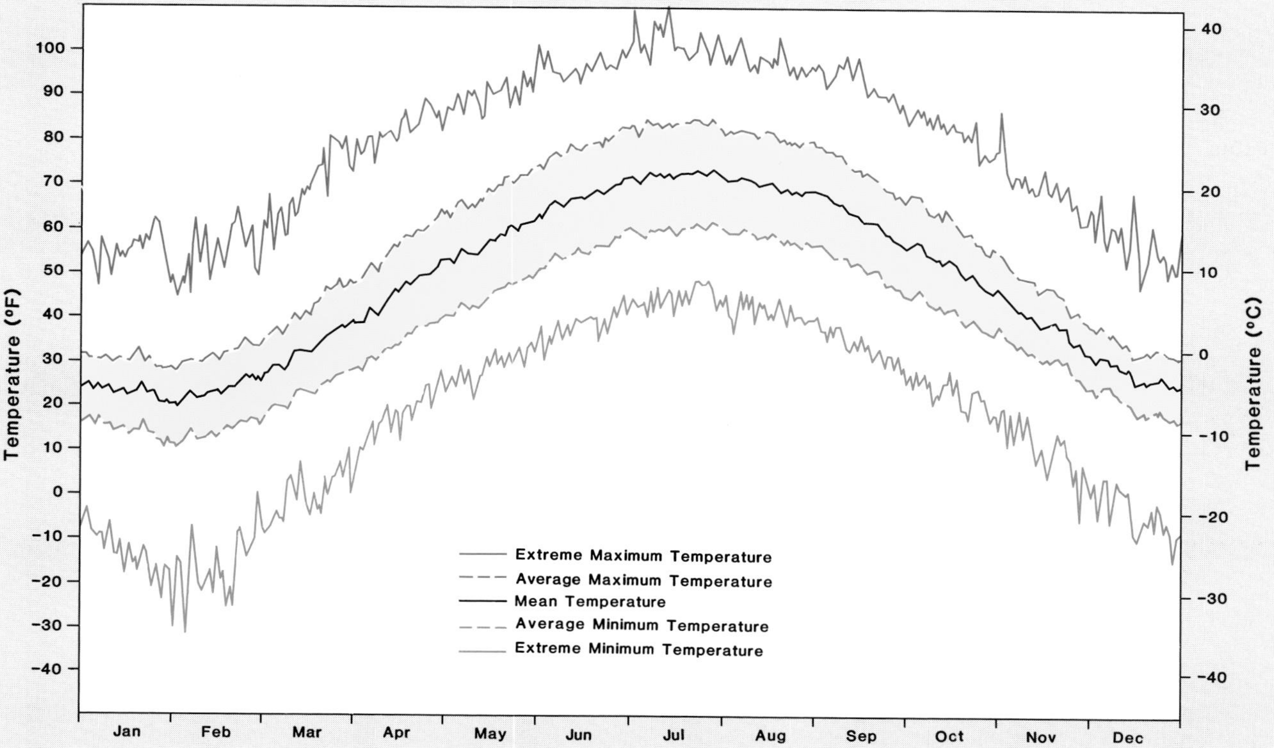


## Battle Creek (1895-1980)

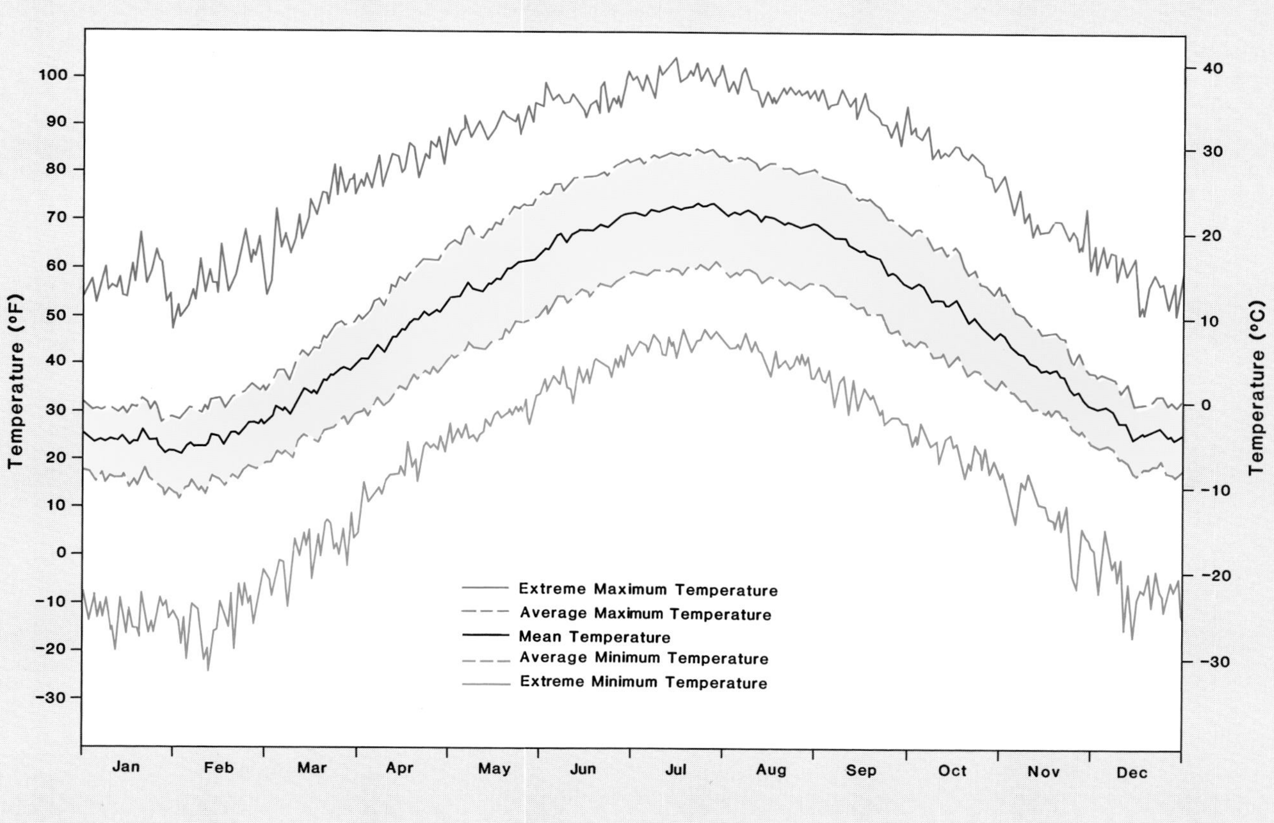

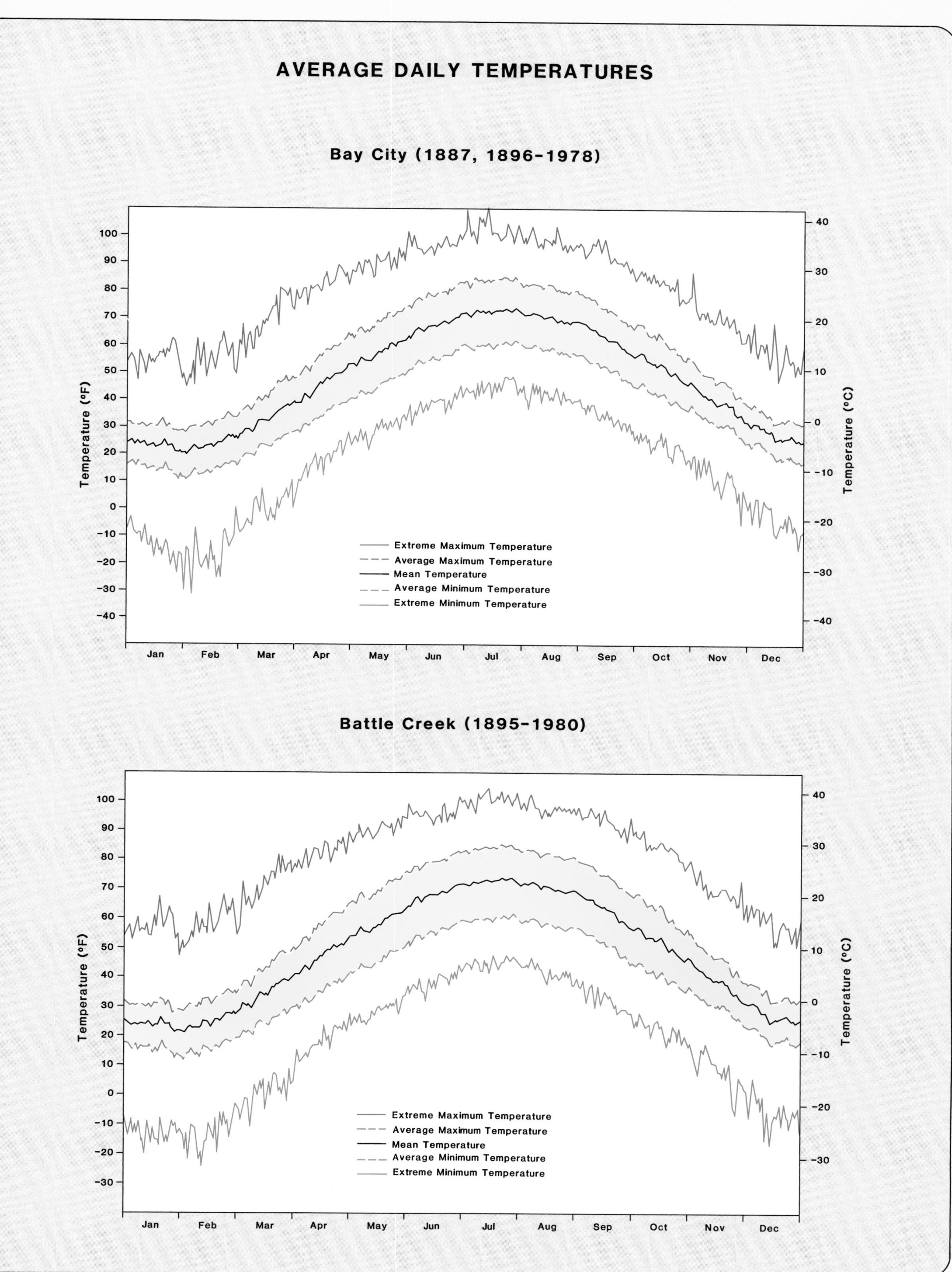

*Source: MDA/Climatology Program*

WMU CARTOGRAPHIC SERVICES DEPARTMENT OF GEOGRAPHY

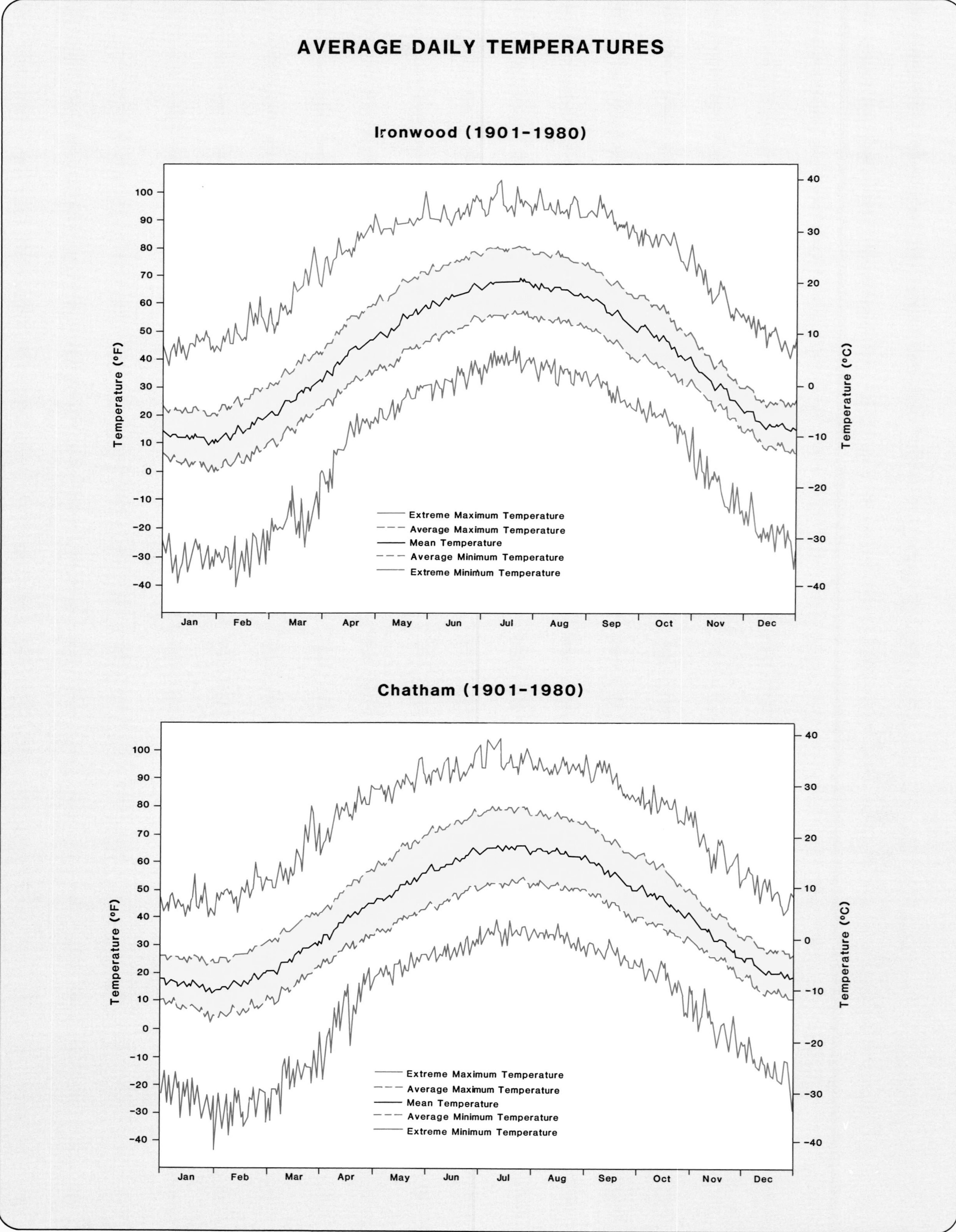


*Source: MDA/Climatology Program*

WMU CARTOGRAPHIC SERVICES DEPARTMENT OF GEOGRAPHY

# AVERAGE DAILY TEMPERATURES

## Cheboygan (1919-1980)

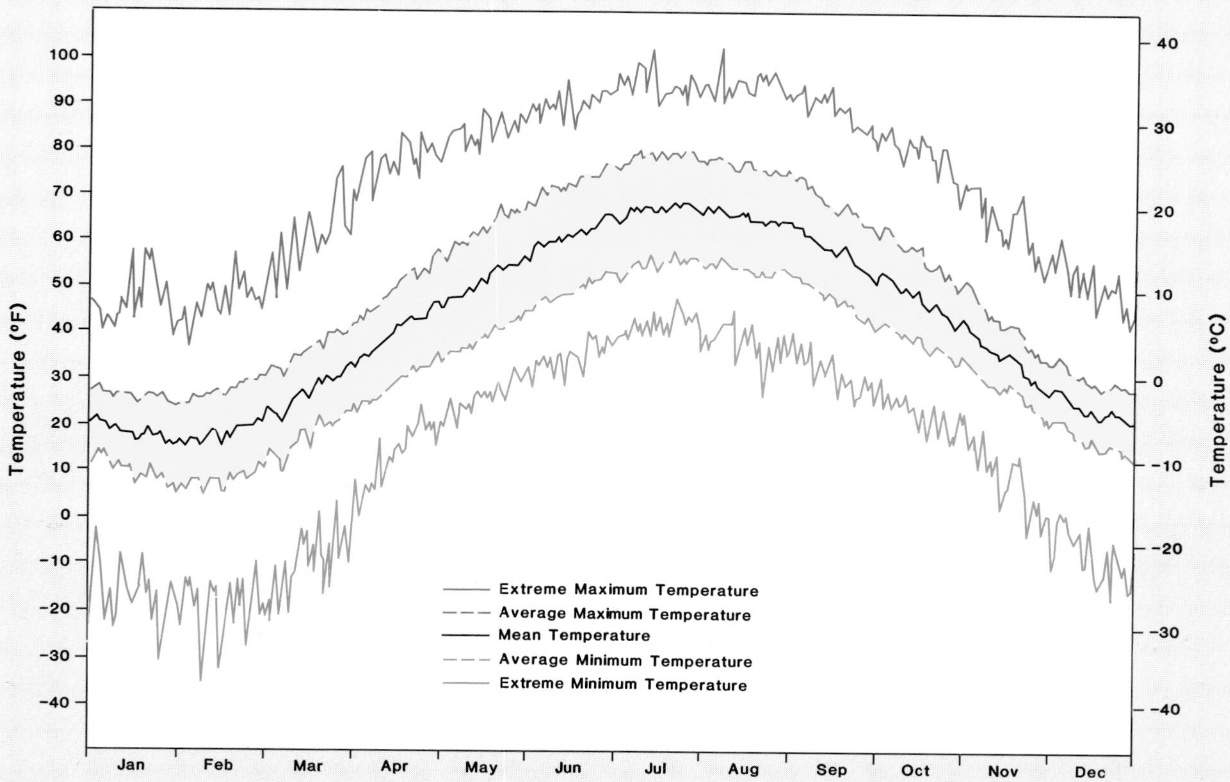


## Traverse City (1931-1980)

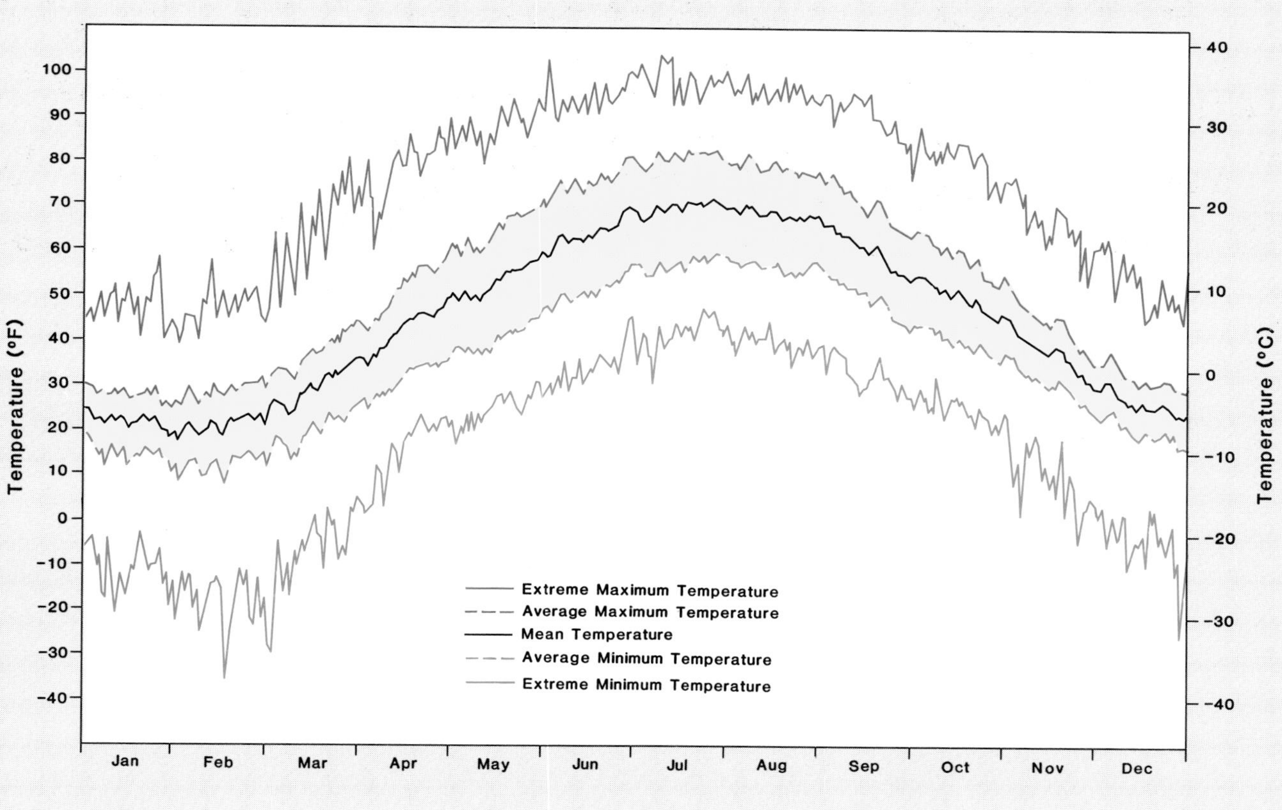

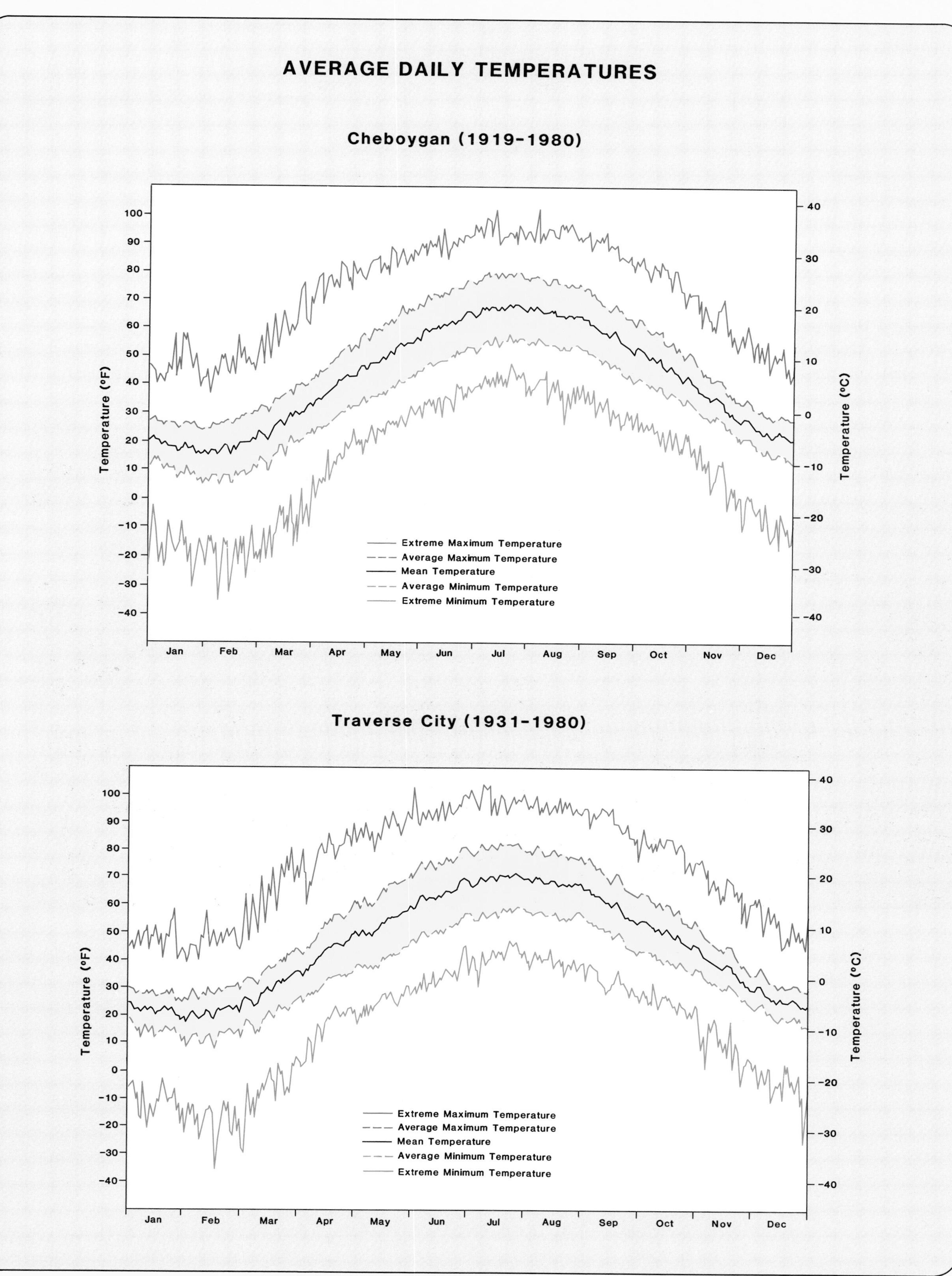

*Source: MDA/Climatology Program*

WMU CARTOGRAPHIC SERVICES DEPARTMENT OF GEOGRAPHY

Morning mist.

# PRECIPITATION

## Introduction

Michigan experiences a humid climate. Annual precipitation normally exceeds evaporation leaving a surplus to provide supportive moisture for vegetation growth and runoff for permanent rivers and streams.

The term "precipitation" includes both liquid and solid forms of moisture derived from the atmosphere. In Michigan, precipitation is primarily rain and snow, but also includes less common forms such as sleet, glaze (freezing rain), and hail. Rain is the dominant form of precipitation throughout the state. The liquid or melted equivalent of snow (actual water content) is used to gauge its contribution to the total precipitation. On the average, the ratio of freshly fallen snow to its liquid equivalent is 10:1, although this can vary greatly, depending on temperature at the time of the snowfall.

Using this ratio, snow would contribute roughly 20% of the annual precipitation in southwestern Michigan, and from 10% to 15% in the southeastern part of the state. In snowbelt areas of the Upper Peninsula and northern Lower Peninsula the contribution of snow to the total precipitation is larger, but usually does not exceed that of rain.

The source of moisture for the majority of Michigan's precipitation is the Gulf of Mexico, although the Great Lakes become secondary sources in winter. The many atmospheric disturbances traversing the state throughout the year cause this moisture to be released as precipitation and contribute to a relatively even precipitation regime. Thus, there are no prominent wet and dry seasons within the state.

Thunderstorms associated with atmospheric disturbances are responsible for much of the warm season precipitation, particularly in the southern portions of the state. As a result, warm season precipitation is more intense, occurs on fewer days, and exhibits an erratic distributional pattern.

During the cool season, precipitation results from low pressure areas, fronts, and their jet stream associations. It is less intense, occurs on more days, and is less erratic in distribution.

Rainshower.

The frequency of flooding is quite low in Michigan; however the greatest likelihood occurs in late winter and early spring when sudden warming may combine with snowmelt to produce large quantities of runoff.

Mild meteorological drought conditions are not uncommon in Michigan, but severe conditions are infrequent and generally of short duration. The normally even distribution of precipitation and higher humidity is helpful in reducing the high demands for crop moisture occurring in other areas of the Midwest.

The data used in the production of all maps in the Precipitation Section of the atlas were observed and recorded by the National Weather Service Network of observers and archived by the MDA/Climatology Program. Computer maps produced by the MDA/Climatology Program were edited and used as a basis for the final atlas maps, with the exception of the annual march of precipitation graphs.

# Average Annual Precipitation

The average annual precipitation in Michigan ranges from 26.00 inches at Sebewaing to 39.14 inches at Niles, although most stations receive around 30 to 34 inches. The range of 50th percentile values, based on the Gamma distribution, is from 25.82 inches to 38.87 inches. Unlike some other climatic features, the average annual precipitation does not exhibit great variability within the state. Existing contrasts are due largely to differences in seasonal snowfall amounts.

The smallest annual precipitation amounts (around 28 inches) occur near Saginaw Bay and on the north and northwest shores of Lake Michigan in the Upper Peninsula. These regions, on the upwind shores of the Great Lakes, receive little lake effect snowfall. Additionally, subsidence of prevailing westerly winds occurs as the air moves from rougher land to smoother water (a downward motion results as winds in the lower layers increase in velocity due to lessened friction). This also contributes to reduced precipitation.

The largest amounts (over 38 inches) occur in the extreme southwest portion of the Lower Peninsula and in the western portion of the Upper Peninsula. In both areas lake effect snowfall is important in the winter and summer precipitation is enhanced by convective activity arriving from areas to the southwest, altitude (in the Upper Peninsula), and in some instances, lake breeze convergence. The southwest Lower Peninsula is also most often affected by moist tropical air masses originating over the Gulf of Mexico and approaching the state on southwesterly surface winds ahead of eastward-moving cyclonic systems. The frequency of this air mass decreases northward because passing cold fronts, associated with these systems, sweep out the moist air masses before they have had sufficient time to reach northern Michigan.

Compared to other parts of the United States, Michigan's annual precipitation is moderate. Portions of the east Gulf Coast exceed 60 inches annually, while coastal areas from northern California to the Canadian border may receive 100 inches or more. On the other hand, much of the western intermontane area is quite dry, with some extreme desert areas averaging less than 5 inches annually.

# Annual March of Precipitation

Monthly averages of precipitation (including water equivalent values for snowfall) are shown for Marquette County Airport, Sault Ste. Marie, Alpena, Houghton Lake, Muskegon, Lansing, and Detroit Metropolitan Airport. In general, the annual march of precipitation shows much less variability than the annual march of temperature. Michigan has a fairly evenly distributed precipitation regime, although cold season amounts are usually somewhat smaller than warm season totals due to the reduced moisture content in the atmosphere.

February is the month of minimum precipitation with amounts less than 2.00 inches for all the stations. The month of maximum precipitation varies. It is April at Muskegon, June at Lansing, Detroit, and Houghton Lake, August at Alpena, and September at Marquette and Sault Ste. Marie. Totals during months of maximum precipitation range between 3 and 4 inches.

In the southern tier of stations a small midsummer dip of precipitation occurs with July totals slightly depressed. Data from earlier periods have shown a more prominent July drop-off of precipitation, a feature which was characteristic of much of the Midwest. A southward shift of the polar front during recent decades has resulted in increased midsummer rainfall and consequently the July decrease of rainfall during the 1951–1980 period is only slightly in evidence.

It is difficult to discern a definitive pattern of monthly precipitation for northern stations, other than larger totals during the warm season. July and August are slightly drier than other summer months at Marquette, and July is slightly drier at Sault Ste. Marie. But June, July, and August have near similar totals at Alpena.

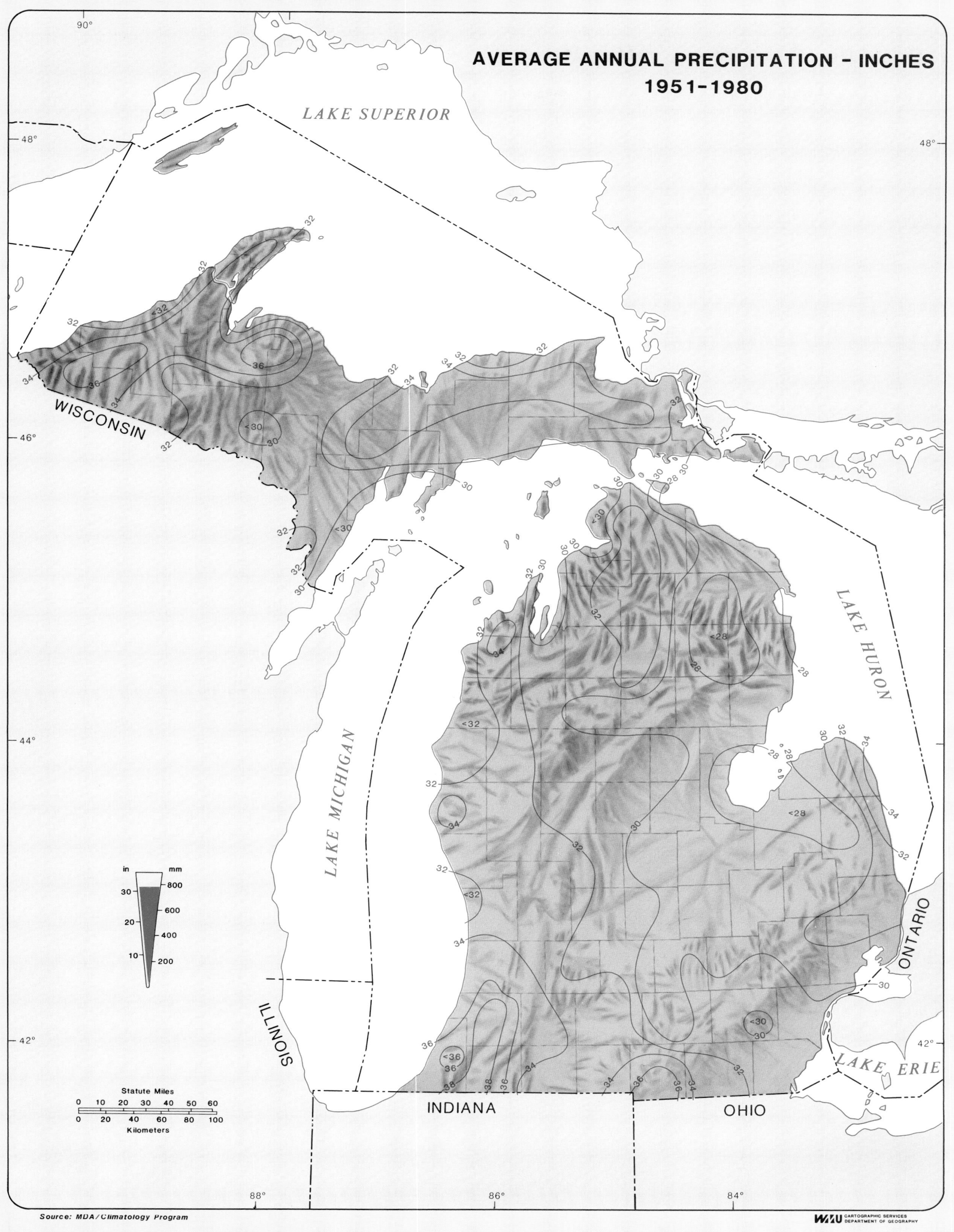


Source: MDA/Climatology Program

WMU CARTOGRAPHIC SERVICES DEPARTMENT OF GEOGRAPHY

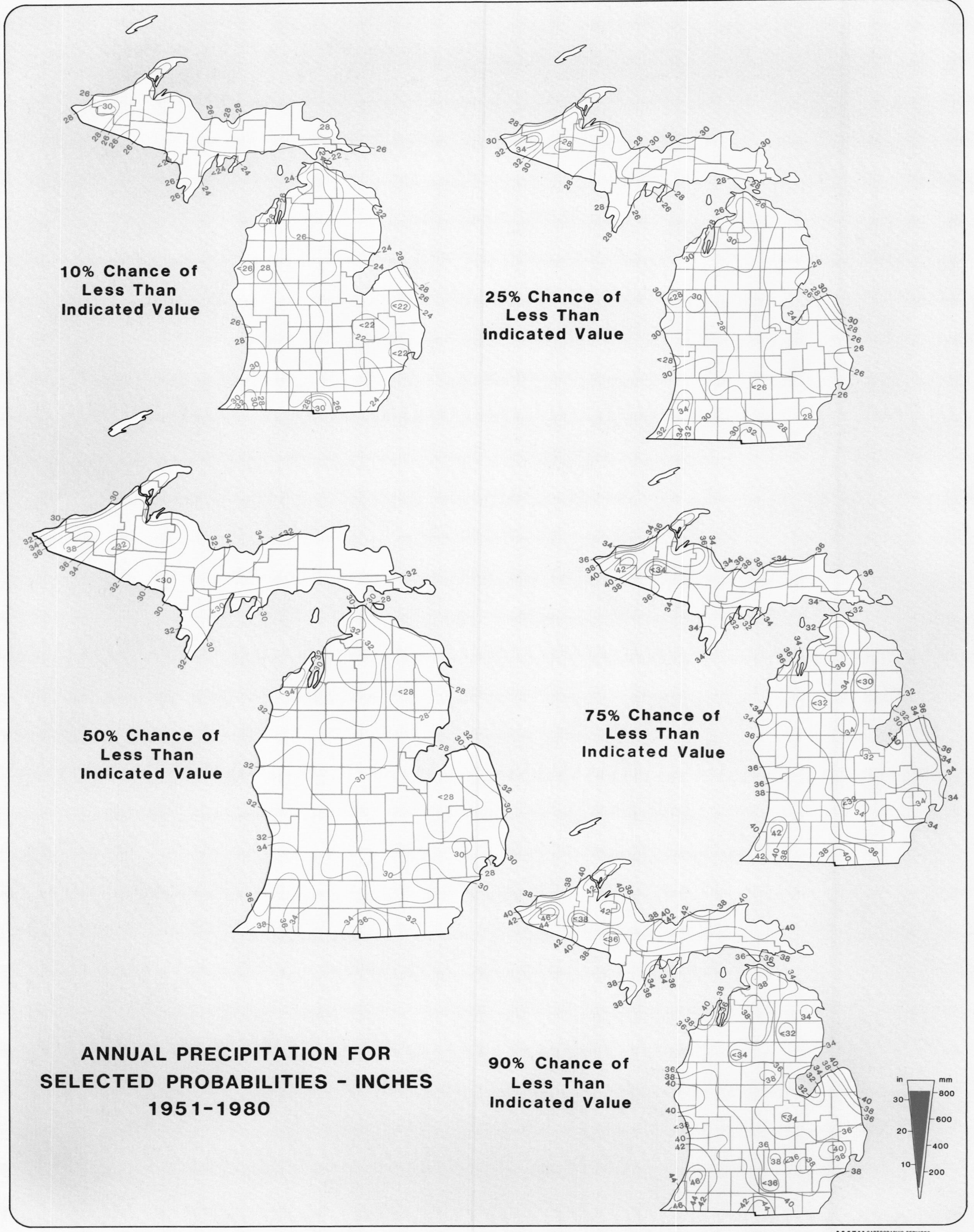


Source: MDA/Climatology Program

WMU CARTOGRAPHIC SERVICES DEPARTMENT OF GEOGRAPHY

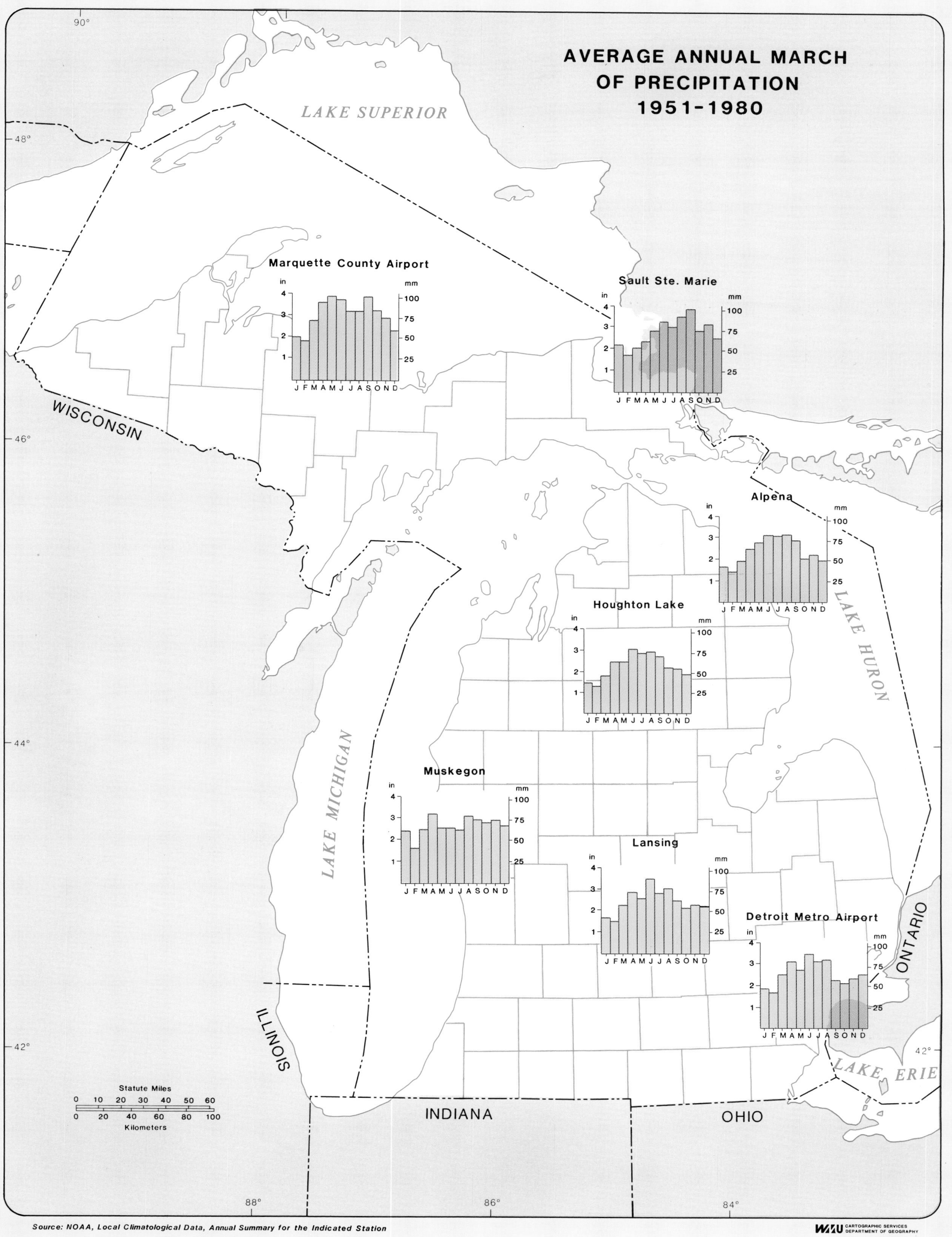


Source: NOAA, Local Climatological Data, Annual Summary for the Indicated Station

Young corn in Berrien County, Michigan.

# Monthly Precipitation

While Michigan's precipitation is essentially evenly distributed with no pronounced wet or dry periods, monthly maps of 50th percentiles of precipitation show that seasonal pulsations occur in regard to the wetter and drier portions of the state. Four more or less distinct spatial patterns can be recognized corresponding to winter, spring, summer, and late summer–early fall.

As cold air can hold less moisture than warm air, January, the coldest month, is relatively dry over much of the state. The 50th percentile (50% probability value) is less than 2 inches at most stations. The role of lake effect snowfall is prominent, and this is a key to the *winter* precipitation pattern in the state. Values in the snowbelt areas exceed 2 inches. The largest 50th percentile value occurs at Houghton FAA Airport, on the Keweenaw Peninsula, with 3.5 inches and the lowest is in the southern interior of the Upper Peninsula with 1.00 inches at Iron Mountain.

February (a cold and short month) is the driest month in the state with 50th percentile values less than 2 inches at all stations except Harbor Beach and Whitefish Point. A number of stations have values of less than 1 inch. With the tapering off of lake effect snowfall in late winter, the winter pattern is less evident than in January.

In March, with increasing warmth, 50th percentile values in the southern portion of the Lower Peninsula exceed 2 inches with some areas near the Indiana border in excess of 2.5 inches. Values in the colder Upper Peninsula and northern Lower Peninsula are smaller, with little evidence of lake effect snow.

April ushers in the *spring* precipitation pattern. 50th percentile values of over 3.5 inches in the extreme southern and southwestern portions contrast with smaller values in the north, particularly along the Lake Superior shoreline where values are less than 2.0 inches. Here lingering ice and cold water suppress vertical motion and restrict precipitation. The extreme south also experiences more thunderstorms (see Thunderstorm Section) and convective activity as temperatures warm toward summer levels. The range of values is from 1.80 inches at Houghton to 3.87 inches at Niles.

In May, as storm tracks swing further northward, 50th percentile values increase in the north while decreasing somewhat in the south. Largest values occur in the western interior portions of the Upper Peninsula.

June begins the *summer* pattern with the largest values in the southwestern Lower Peninsula and western Upper Peninsula. Areas along the Lake Michigan shoreline in the Lower Peninsula evidence some lake suppression, with less than 2.5 inches.

July continues the summer pattern with the largest values (greater than 3.5 inches) in extreme southern Michigan and portions of the western Upper Peninsula. The higher elevation interior areas of the northern Lower Peninsula and the western Upper Peninsula have values over 3.00 inches. The range of values is from 3.77 inches at Ironwood to 1.98 inches at Ludington.

August begins the *late summer–early fall* regime, when 50th percentile values in the Upper Peninsula are larger than those in much of the Lower Peninsula. The majority of the Upper Peninsula can expect more than 3.00 inches with Ironwood's value being 4.16 inches. Values in the Lower Peninsula are generally less than in July, with the area of greater than 3.00 inches located in the west-central portion.

September continues the pattern established in August. With storm tracks shifting southward from Canada, much of the northern Lower Peninsula and the Upper Peninsula have values greater than 3.00 inches, with some areas more than 3.5 inches. Amounts in the southern Lower Peninsula are generally less than 3.0 inches, and in most southeast counties less than 2.5 inches.

In October, differences between northern and southern parts of the state are small, and only Niles (3.10") and Munising (3.07") have 50th percentile values more than 3.00 inches. The southeast part of the Lower Peninsula is still the driest part of the state, with values generally less than 2.00 inches. The lowest values are 1.65 inches at Escanaba and 1.67 inches at Mt. Clemens.

November's map begins to show evidence of lake effect snowfall, particularly in the Upper Peninsula, where values may exceed 3.00 inches. The southern interior portion of the Upper Peninsula and the southeastern part of the Lower Peninsula remain drier, with less than 2.5 inches.

The month of December heralds the return of the *winter* pattern, with a strong impact of lake effect snowfall. The eastern portion of the Lower Peninsula and the north shore of Lake Michigan in the Upper Peninsula have values of less than 2.00 inches, while Whitefish Point in the eastern Upper Peninsula and Bloomingdale in southwestern Lower Michigan are the only stations with values of 3.00 inches or more.

Cumulus cloud building in humid air.

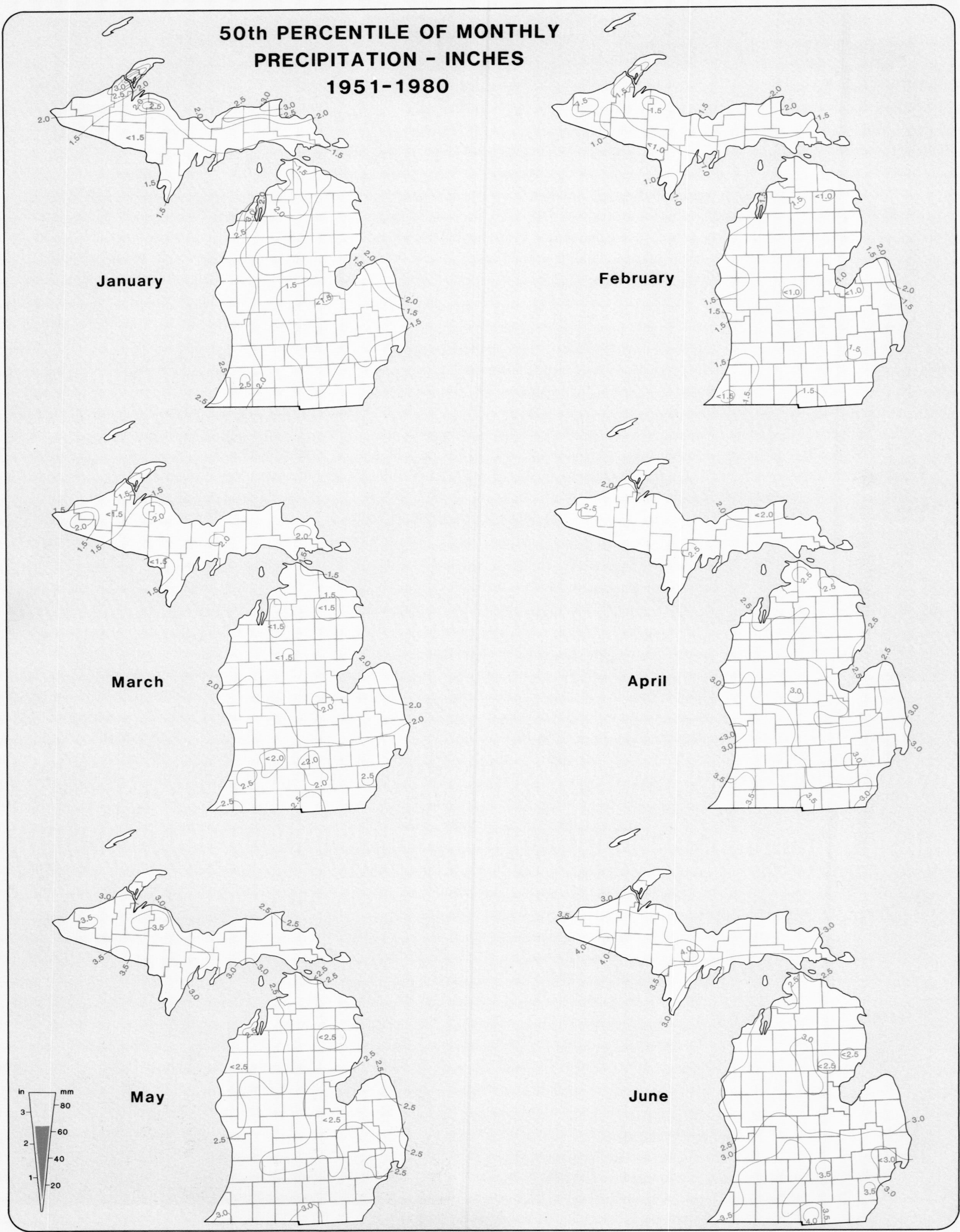


Source: MDA/Climatology Program

WMU CARTOGRAPHIC SERVICES DEPARTMENT OF GEOGRAPHY

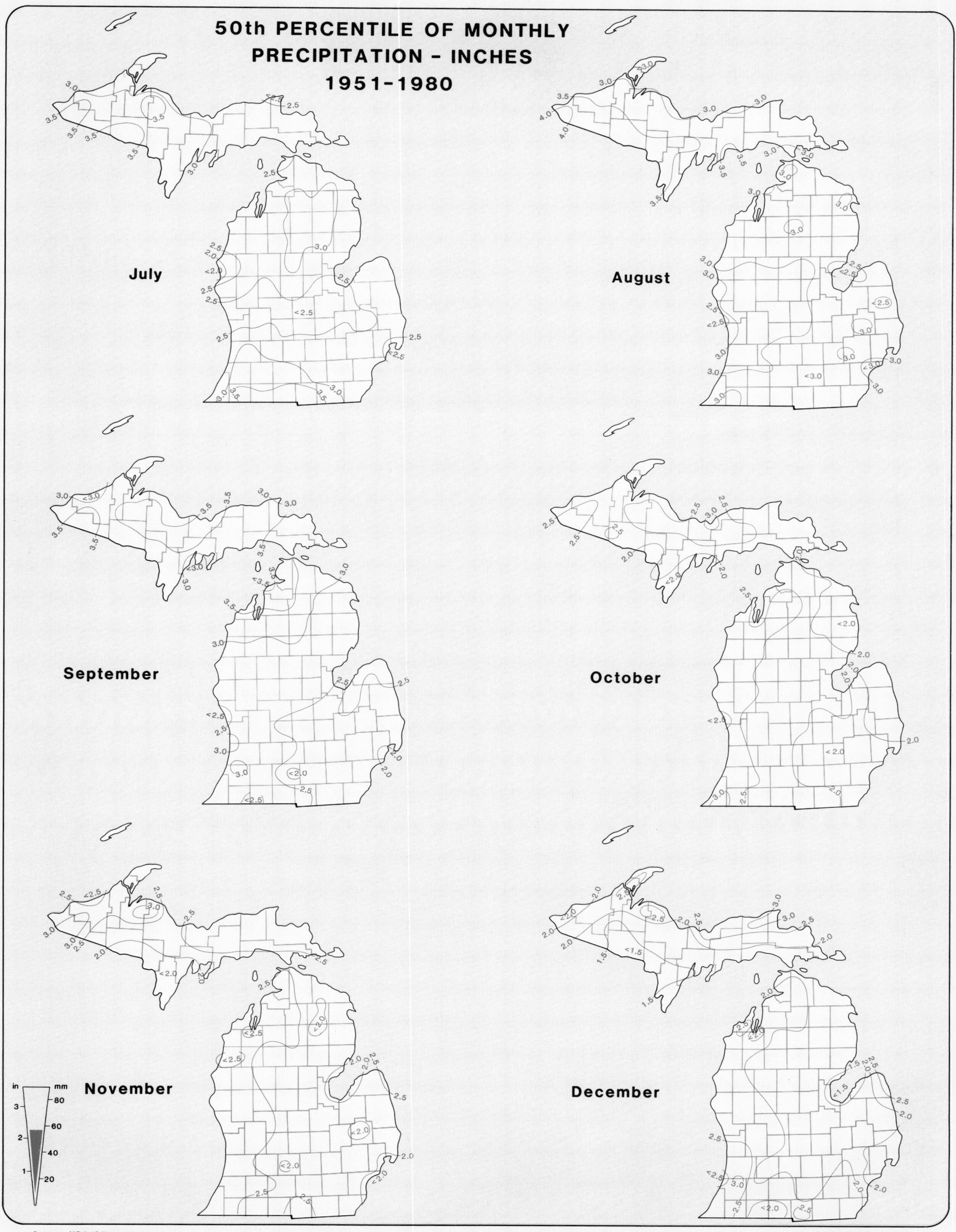


Source: MDA/Climatology Program

WMU CARTOGRAPHIC SERVICES DEPARTMENT OF GEOGRAPHY

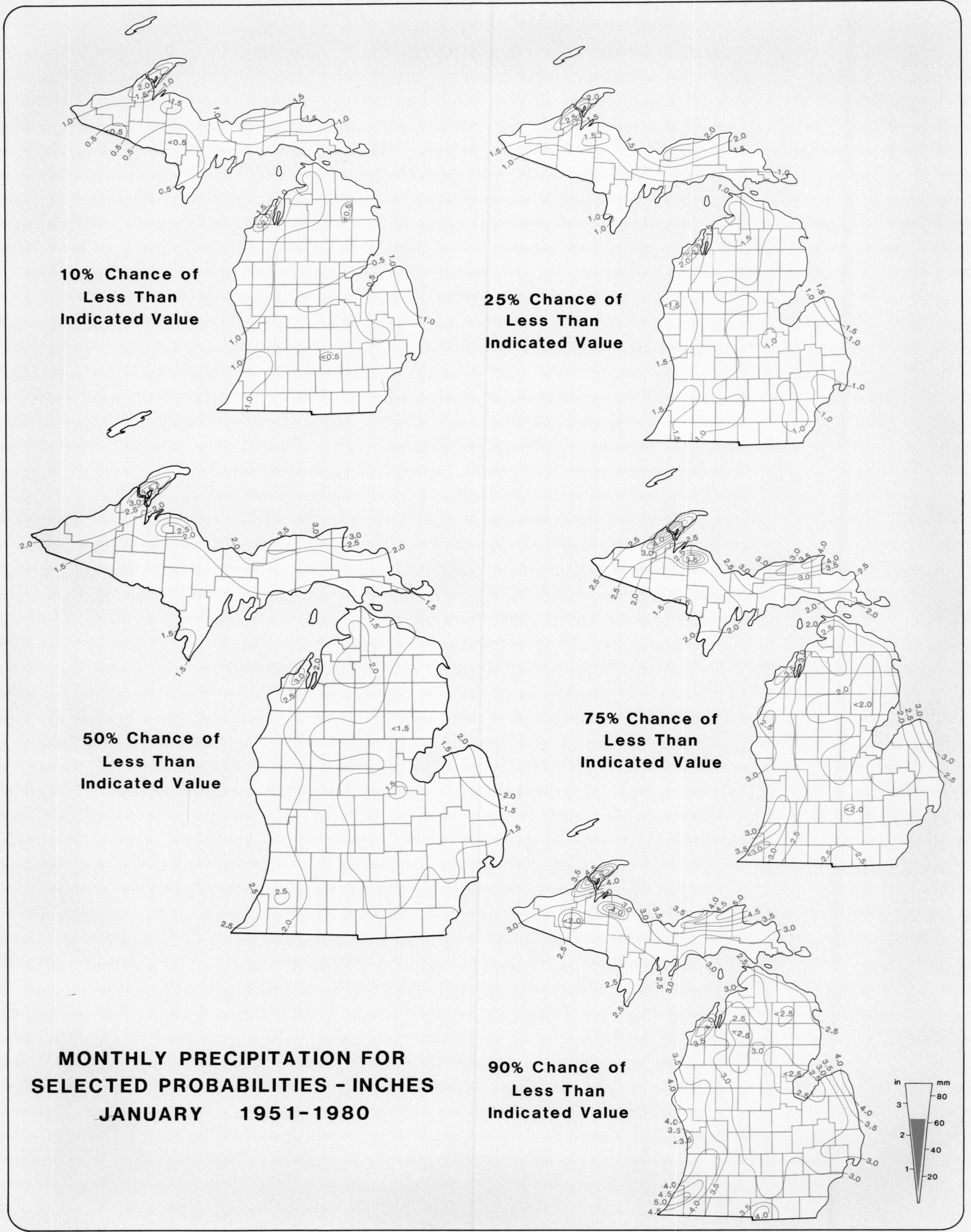


MONTHLY PRECIPITATION FOR SELECTED PROBABILITIES - INCHES
JANUARY 1951-1980

Source: MDA/Climatology Program

WMU CARTOGRAPHIC SERVICES DEPARTMENT OF GEOGRAPHY

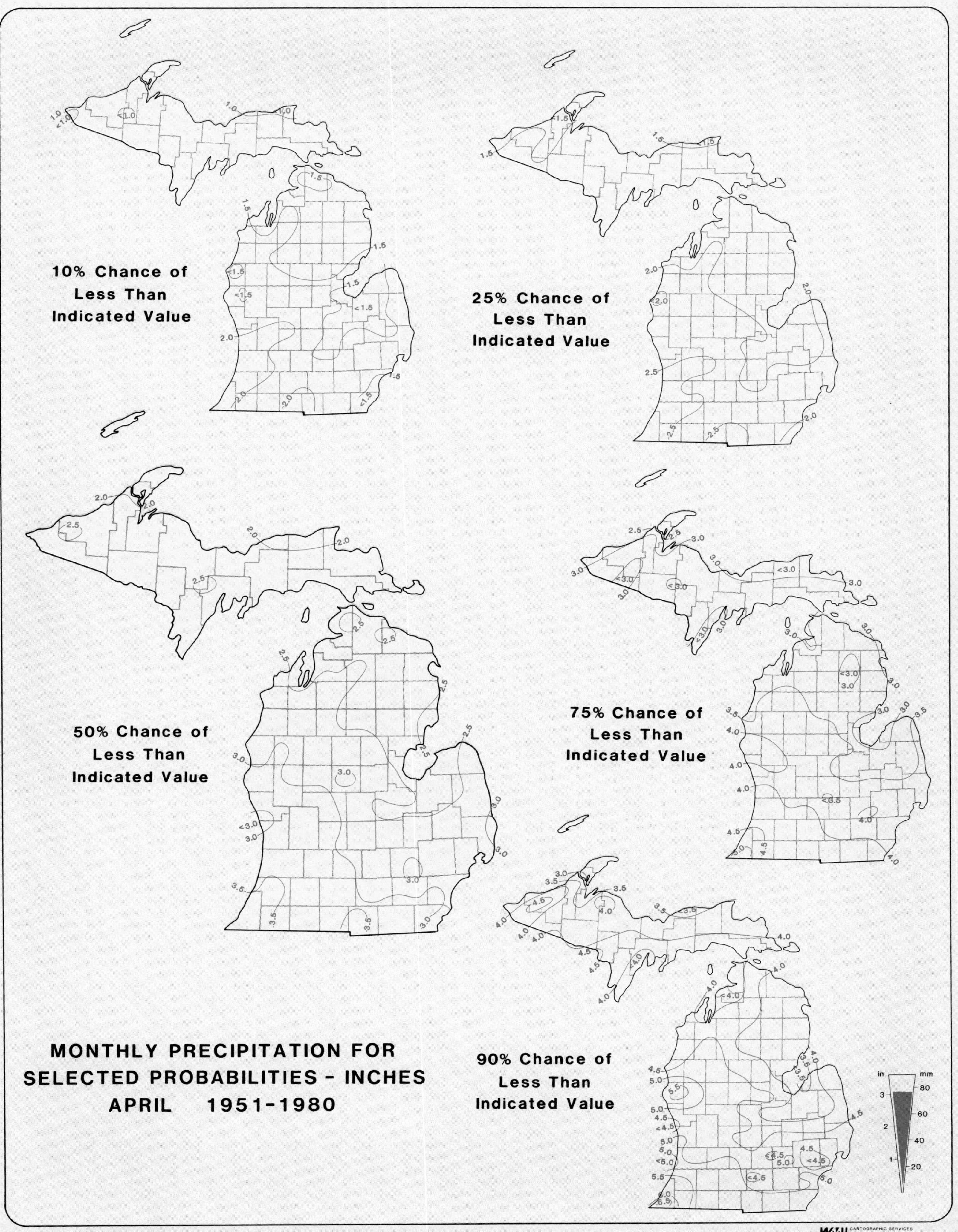
10% Chance of
Less Than
Indicated Value
25% Chance of
Less Than
Indicated Value
50% Chance of
Less Than
Indicated Value
75% Chance of
Less Than
Indicated Value
90% Chance of
Less Than
Indicated Value
MONTHLY PRECIPITATION FOR
SELECTED PROBABILITIES - INCHES
APRIL 1951-1980
in
mm
Source: MDA/Climatology Program
WMU CARTOGRAPHIC SERVICES
DEPARTMENT OF GEOGRAPHY

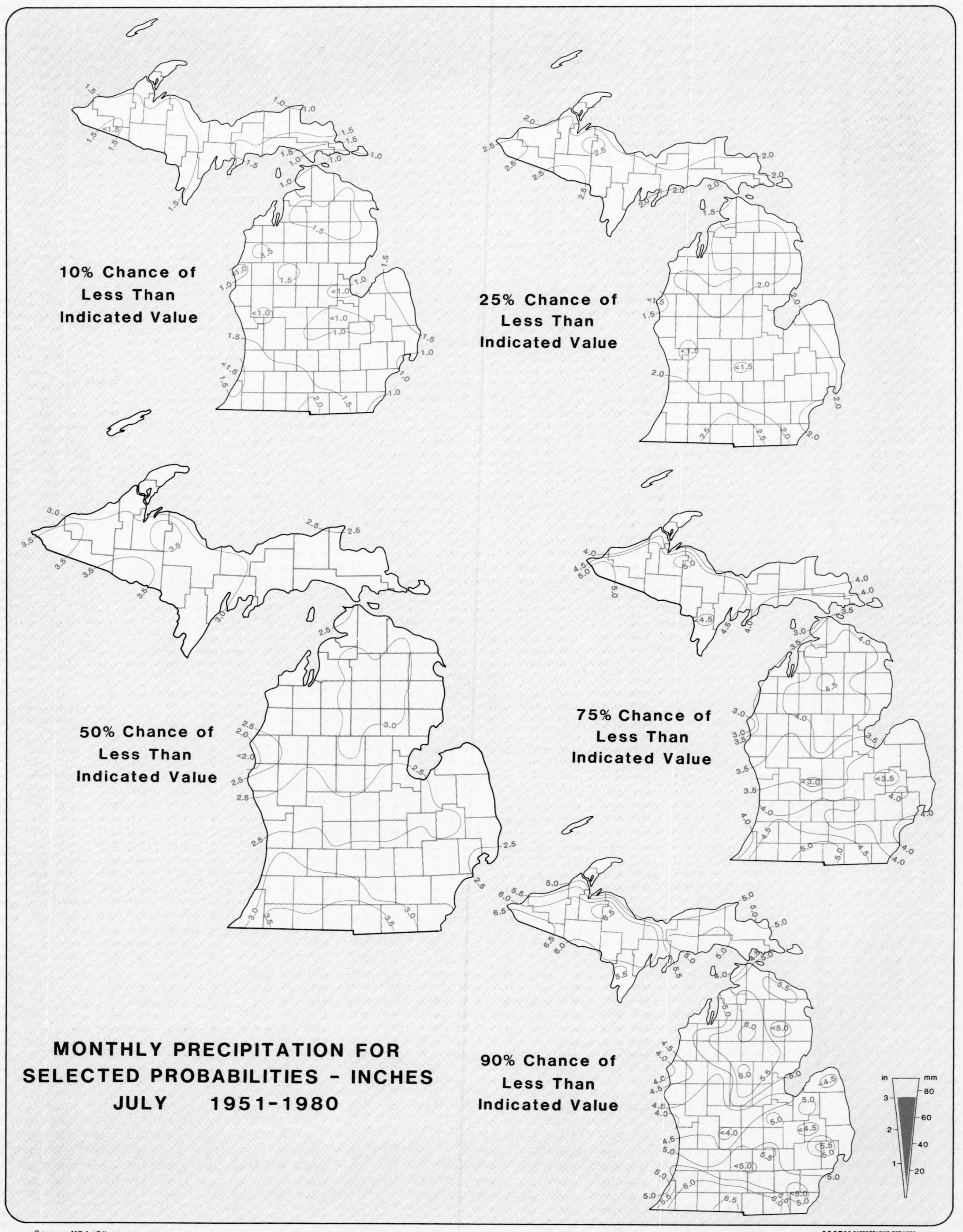


Source: MDA/Climatology Program

WMU CARTOGRAPHIC SERVICES DEPARTMENT OF GEOGRAPHY

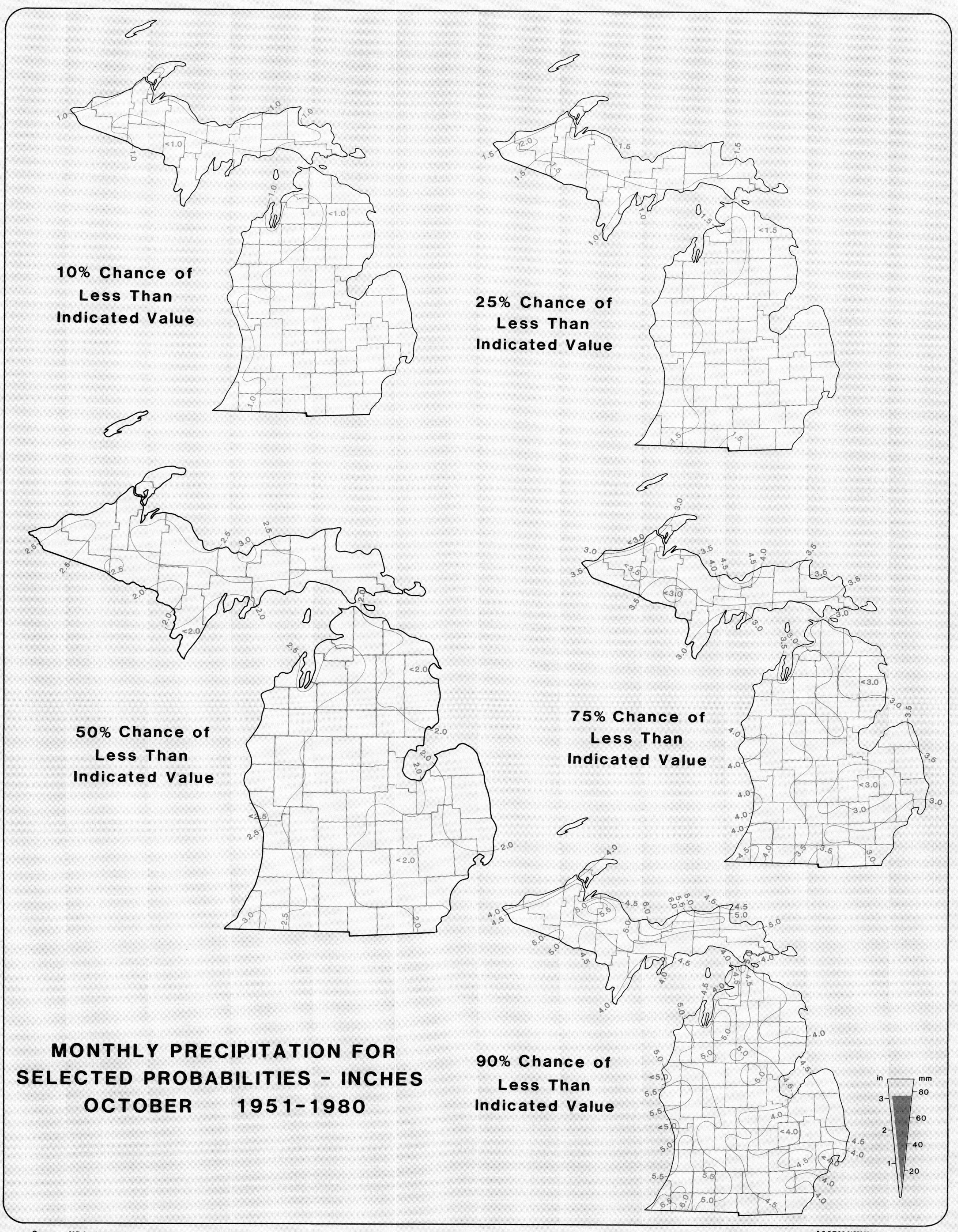


Source: MDA/Climatology Program

WMU CARTOGRAPHIC SERVICES DEPARTMENT OF GEOGRAPHY

# Extreme Maximum Monthly Precipitation

The excessively wet month of September 1986 makes a strong imprint on the map of Michigan, dominating the map pattern in the Lower Peninsula. During that very rainy month, totals exceeded 18 inches in the central portion of the Lower Peninsula. Elsewhere, the records were set on numerous dates and the amounts vary. The values are greater than 8 inches in all parts of the state with only two exceptions, Dearborn (7.78 inches) and Big Bay (7.91 inches). Both of these stations have relatively short periods of record.

This section, like the extreme temperature section, is not restricted to a fixed time frame. Each station extreme is for whatever the complete record is for that station. Thus, some records go back to the third quarter of the 1800s while others start in the mid-1900s.

The state record through 1987 is 19.26 inches at Edmore in September 1986. Prior to that month, the record had been 16.24 inches at Battle Creek in June 1883. For September 1986, nine long-term and ten short-term stations equaled or exceeded the previous state record with four of the stations recording 19.00 inches or more for the month.

# Average Number of Days with Precipitation of .10 Inches or More

## Annual

The average annual number of days in Michigan with precipitation of .10 inches or more ranges from 64 at Cheboygan to 92 at Bergland Dam. The variation within the state is largely a response to lake effect snowfall. In snowbelt areas, daily liquid equivalents resulting from lake effect snowfall may frequently exceed .10 inches. In most of the state's snowbelt areas the number of days of .10 inches or more exceeds 85 in the Upper Peninsula and 80 in the Lower Peninsula. Outside the snowbelts, the total is usually less than 70. Thus, the lake effect increases the annual number of days by 20 to 30% compared to the inland areas.

## Monthly

The largest differences within the state occur in the winter when lake effect snowfall is at its peak and the winter precipitation pattern is well established. In January, the snowbelt areas of the Upper Peninsula average over 11 days, while over half the Lower Peninsula averages less than 5. In February, the pattern is similar to January with the number of days and range of days reduced.

In spring, as the winter precipitation pattern disappears, values increase in the south and decrease in the north. By April, the spring precipitation pattern has established itself with over 8 days in the south and only 5 or 6 in the Upper Peninsula.

By May, values begin to increase in the western interior of the Upper Peninsula with some areas having 8 days. Monthly maps through August show minimal contrasts within the state.

In September, with the late summer–early fall precipitation pattern established, the Upper Peninsula averages more days (generally over 8) than the southern part of the state (5 or 6). By October, the number of days decreases in the Upper Peninsula and increases in the Lower Peninsula, so that contrasts within the state are again small. By November, the winter precipitation regime begins to reestablish itself in the Upper Peninsula where some portions are in the 8 to 9 range, and by December it is dominant throughout the state, with regions having the largest values in the 10 to 11 range.

Lake Michigan.

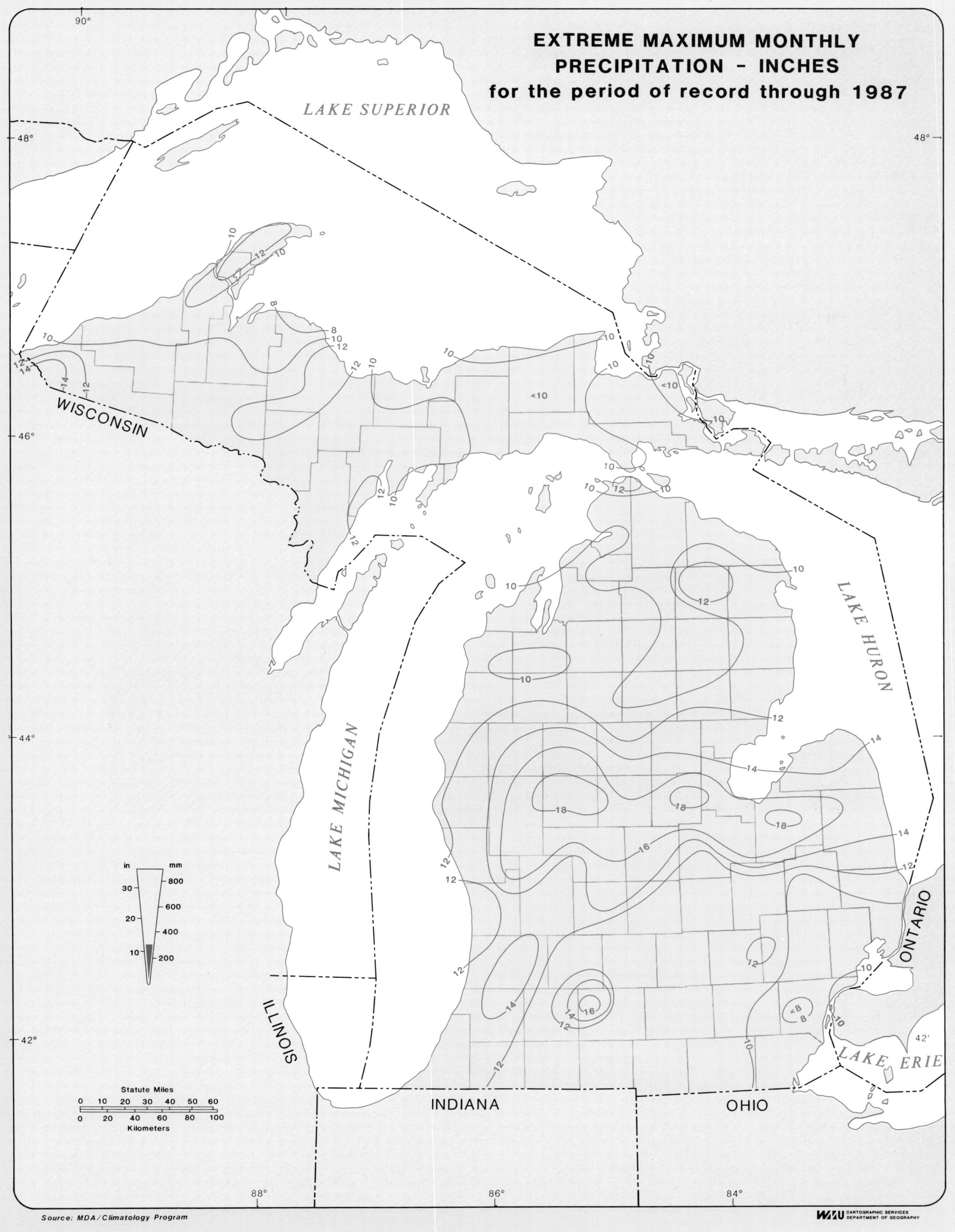
EXTREME MAXIMUM MONTHLY
PRECIPITATION - INCHES
for the period of record through 1987
LAKE SUPERIOR
LAKE MICHIGAN
LAKE HURON
LAKE ERIE
WISCONSIN
ILLINOIS
INDIANA
OHIO
ONTARIO
90°
88°
86°
84°
48°
46°
44°
42°
in
mm
Statute Miles
Kilometers
Source: MDA/Climatology Program
WMU CARTOGRAPHIC SERVICES DEPARTMENT OF GEOGRAPHY

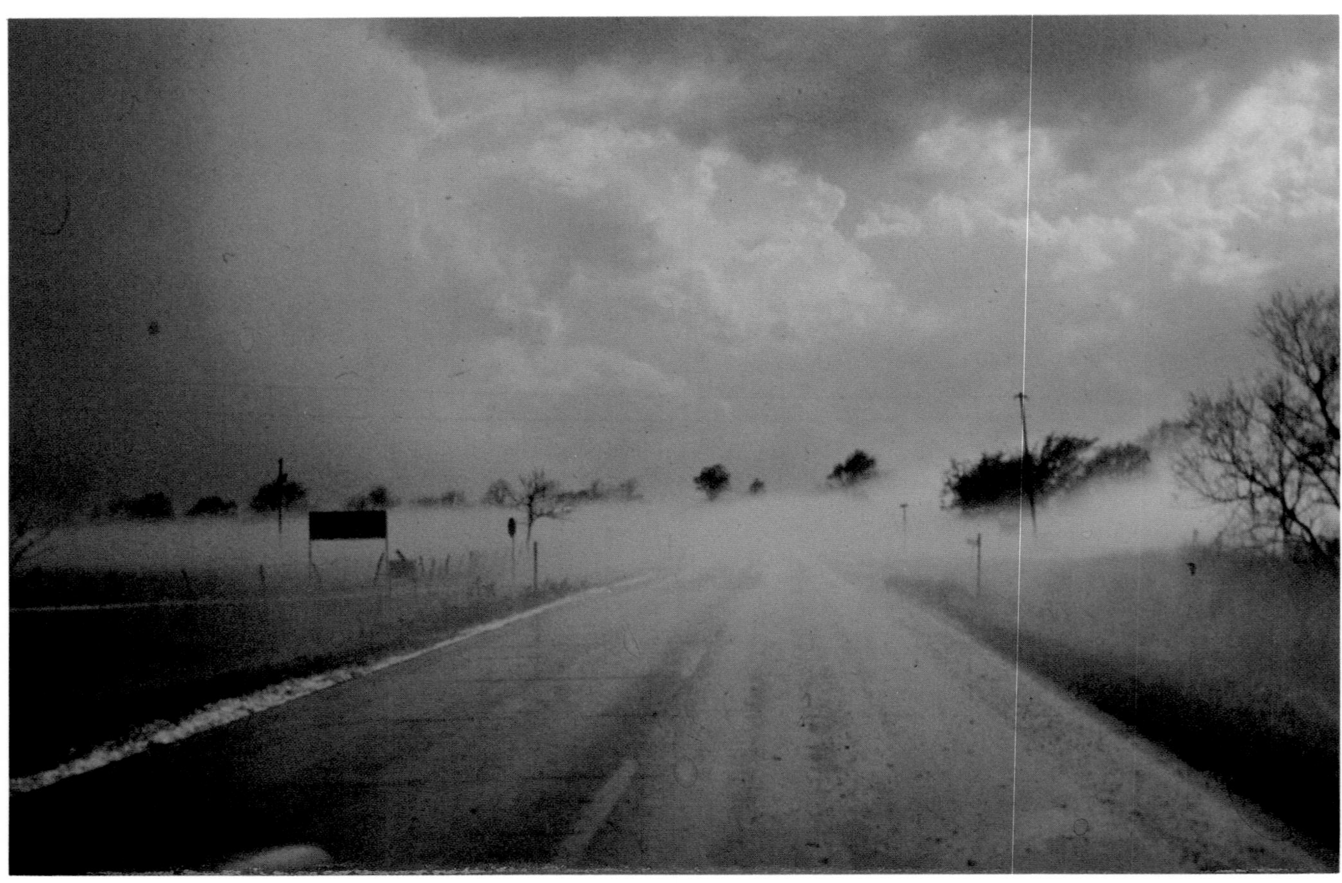

Fog on highway.

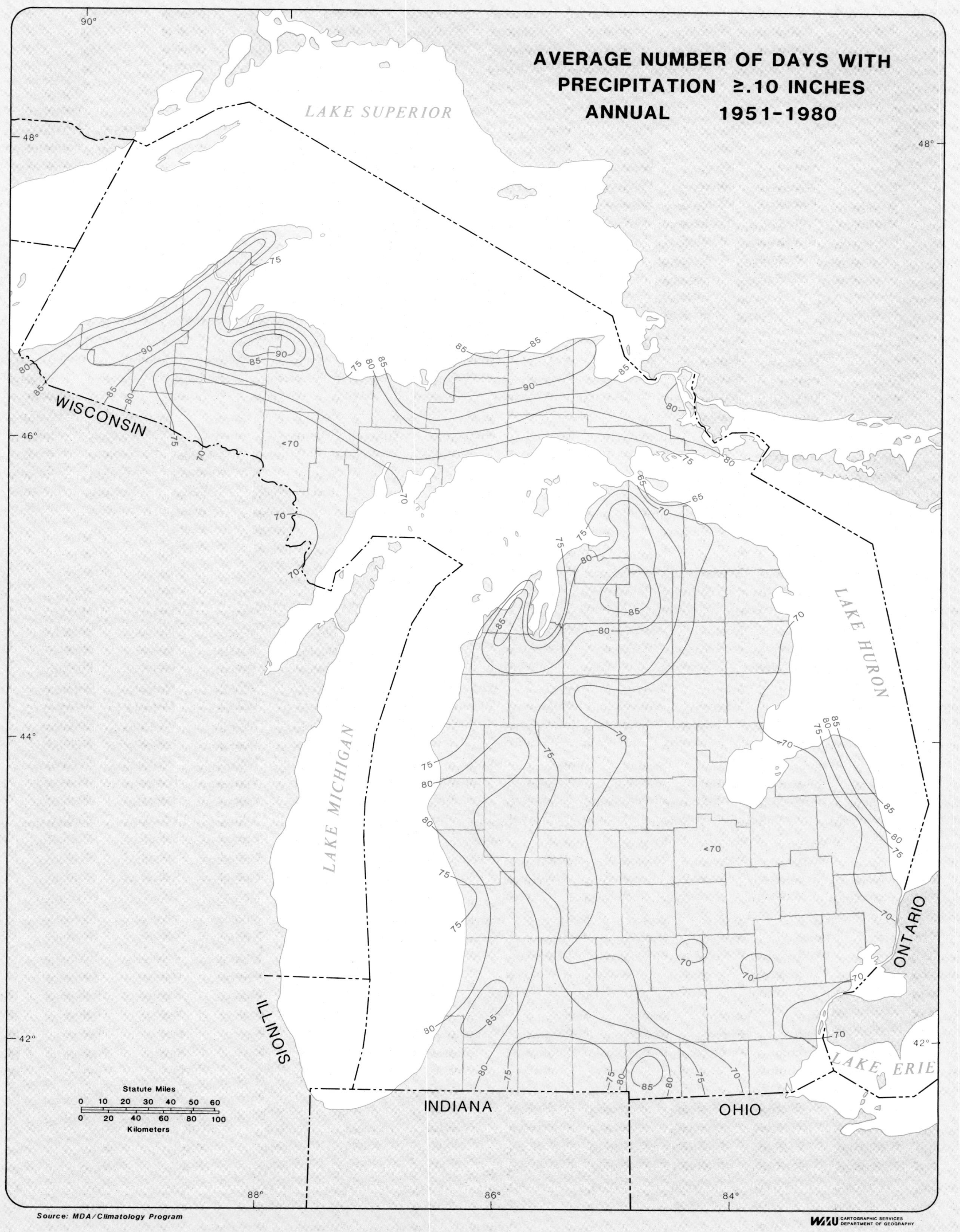


Source: MDA/Climatology Program

WMU CARTOGRAPHIC SERVICES DEPARTMENT OF GEOGRAPHY

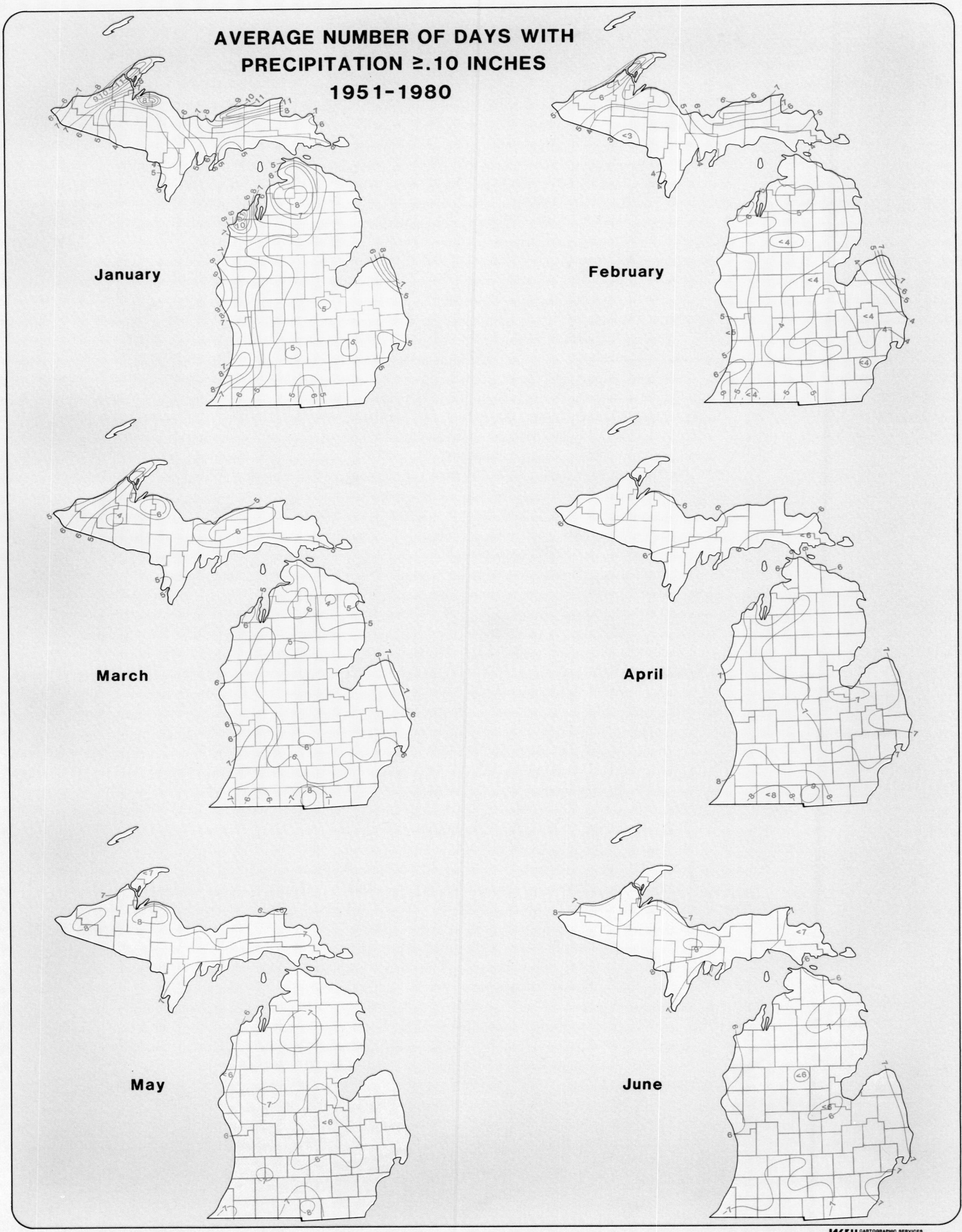


Source: MDA/Climatology Program

WMU CARTOGRAPHIC SERVICES DEPARTMENT OF GEOGRAPHY

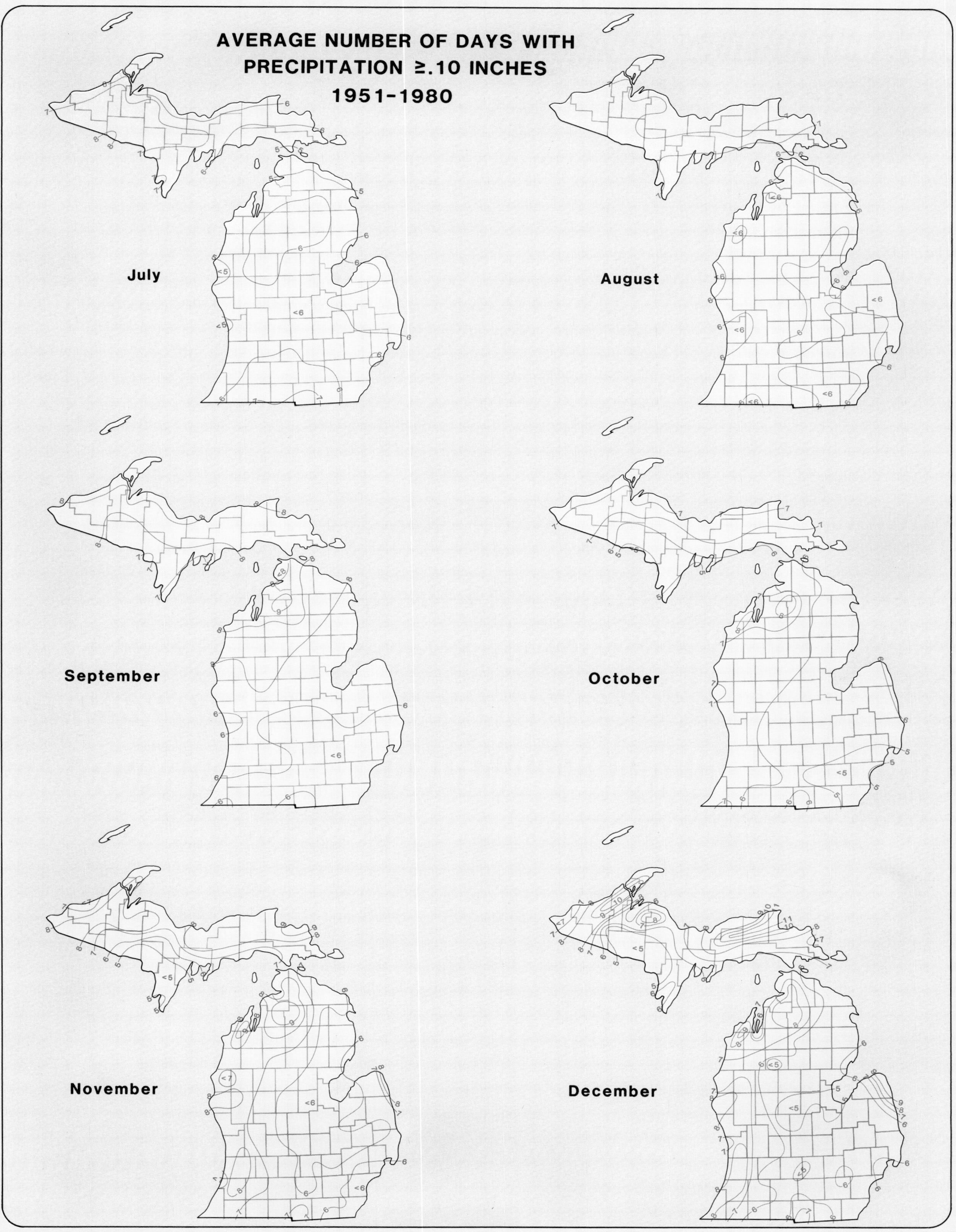


Source: MDA/Climatology Program

WMU CARTOGRAPHIC SERVICES DEPARTMENT OF GEOGRAPHY

# Average Number of Days with Precipitation of .25 Inches or More

### Annual

The annual pattern ranges from a maximum of 47 at a number of southern Michigan stations to a minimum of 34 at Mio. The differences shown by the map are smaller than those shown by the .10 inches map as the impact of lake effect snowfall is not as pronounced. This is because the water content in daily lake effect snowfall amounts may frequently exceed .10 inches in snowbelt areas, but less frequently exceed .25 inches.

### Monthly

In general, similar monthly patterns as occurring with the .10 inches maps also occur with the .25 inches maps, although the contrasts within the state are less distinct. The snowbelt areas have the most days in the winter, particularly in the Upper Peninsula. In March and April, values increase in the southern portion of the Lower Peninsula, and decrease toward the north. In May, June, July, and August differences within the state are small. In September, the Upper Peninsula has the largest values. In November and December, the winter lake effect pattern returns.

### Seasonal (October-March, April-September)

The selected seasonal maps, April-September for the warm season and October-March for the cold season, serve to illustrate the more significant influence of lake effect snow on the cold season totals. Cold season values range from a maximum of 23 days at Whitefish Point, Benton Harbor, and Harbor Beach to a minimum of 12 days at a number of stations in the interior of the western Upper Peninsula and in the north central Lower Peninsula. Warm season values range from 27 days at stations in the interior western Upper Peninsula to 21 days at Mt. Clemens and St. Charles in the Lower Peninsula.

# Average Number of Days with Precipitation of .50 Inches or More

### Annual

The impact of lake effect snowfall on the map pattern is lessened as the majority of daily lake effect snows have water content of less than .50 inches. Some impact of lake effect snow is still apparent, however. The largest number of days occurs in the extreme southwestern portion of the state (Dowagiac 24), while the smallest values occur at Alpena and Mio Hydro Plant with 15. Most of the Upper Peninsula is in the 18 to 21 range while most of the stations in the Lower Peninsula range from 16 to 23.

### Monthly

Most monthly maps show indefinite patterns, with only small contrasts throughout the state. The effect of lake effect snowfall in the winter is not as clearly discernible. There is some indication of the spring precipitation regime with larger values in the southern part of the state. A hint of the late summer–fall pattern shows on the September map.

### Seasonal (April-September, October-March)

The warm season and the cold season maps help to illustrate the effects of thunderstorms producing .50 inch or greater days in the summer. The warm season values range from 15 at Watersmeet to 10 at Mt. Clemens. The cold season values range from 9 at three locations in the extreme south central and southwest Lower Peninsula to several stations with less than 5 in the west central Upper Peninsula.

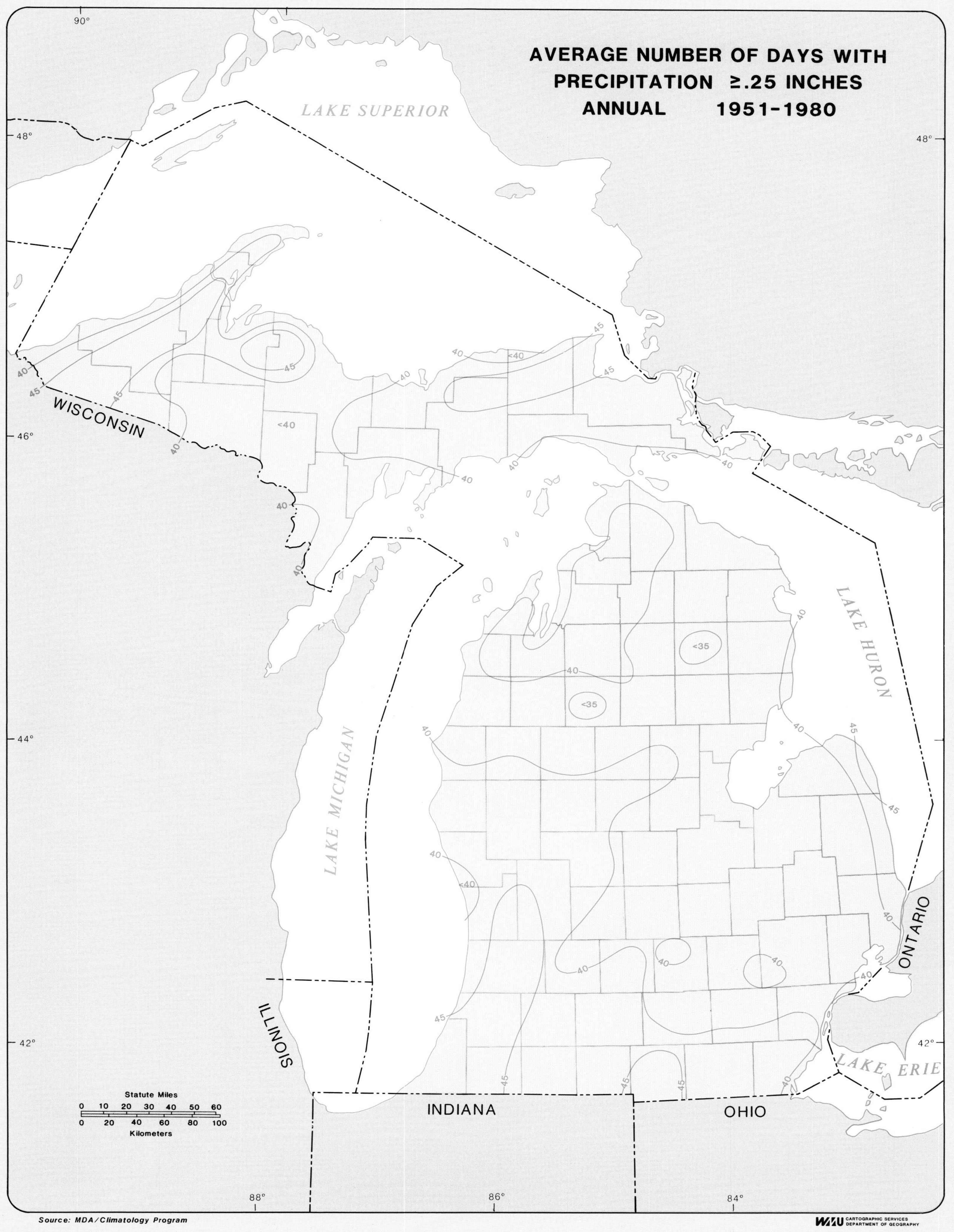


Source: MDA/Climatology Program

WMU CARTOGRAPHIC SERVICES DEPARTMENT OF GEOGRAPHY

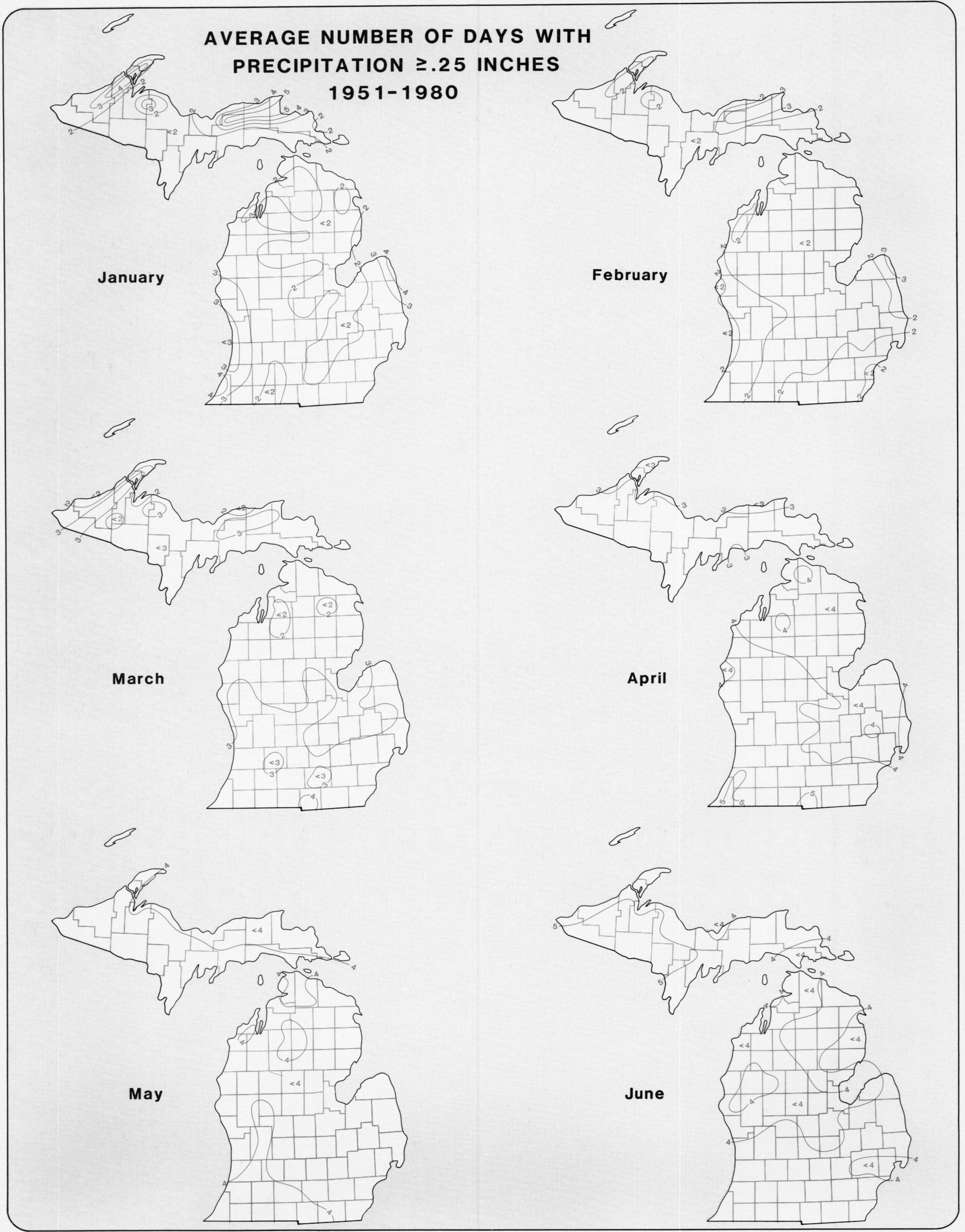


Source: MDA/Climatology Program

WMU CARTOGRAPHIC SERVICES DEPARTMENT OF GEOGRAPHY

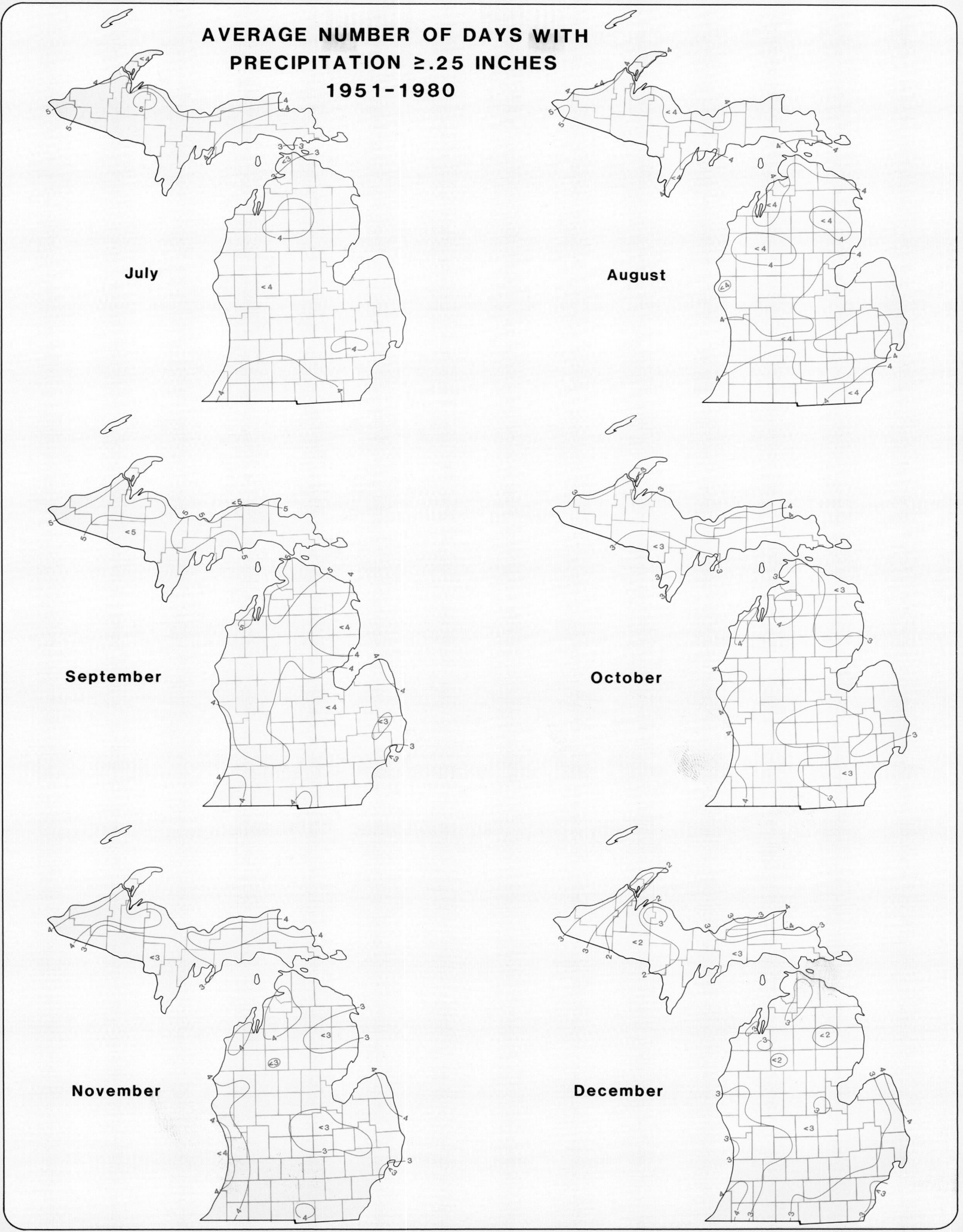


Source: MDA/Climatology Program

WMU CARTOGRAPHIC SERVICES DEPARTMENT OF GEOGRAPHY

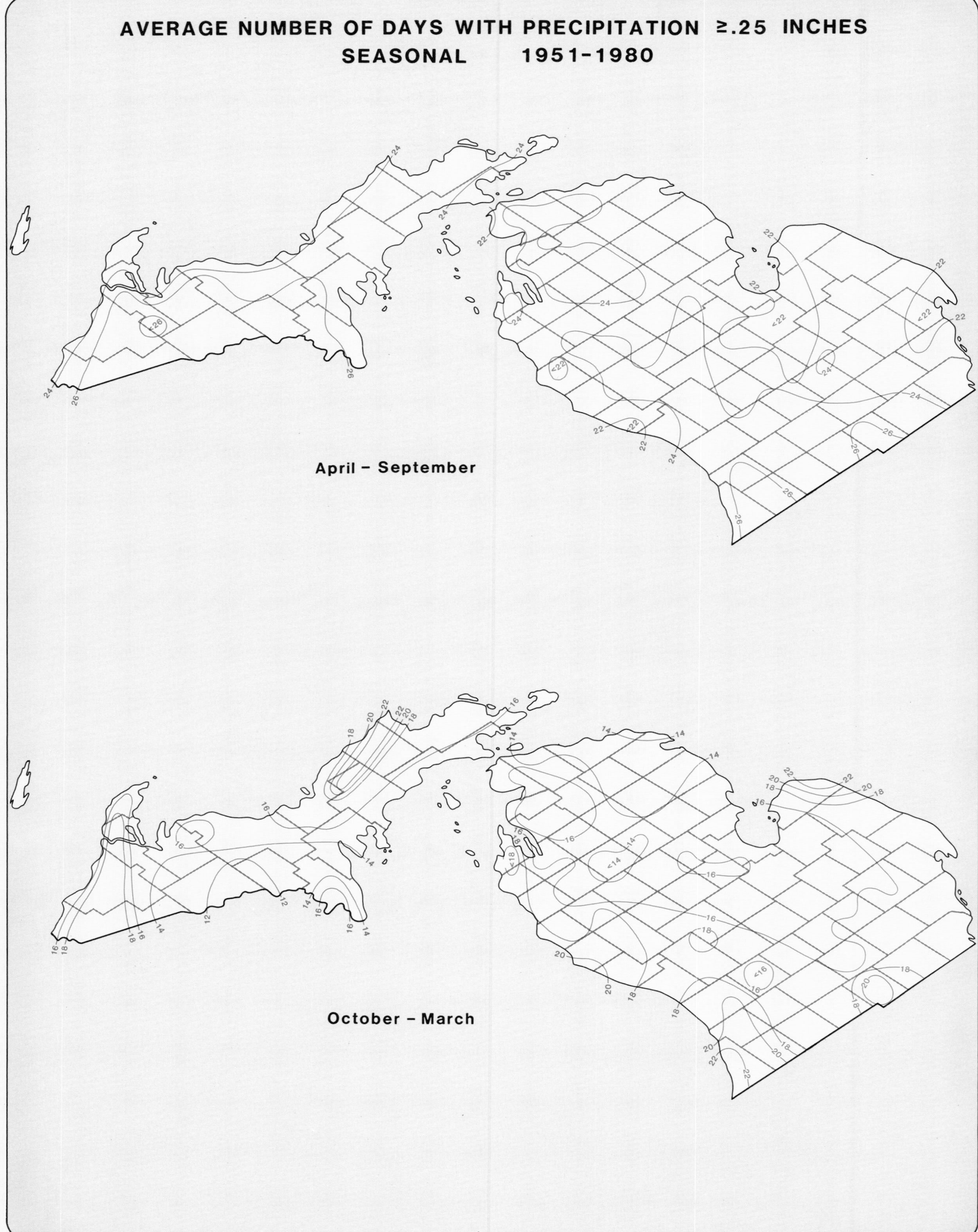


Source: MDA/Climatology Program

WMU CARTOGRAPHIC SERVICES DEPARTMENT OF GEOGRAPHY

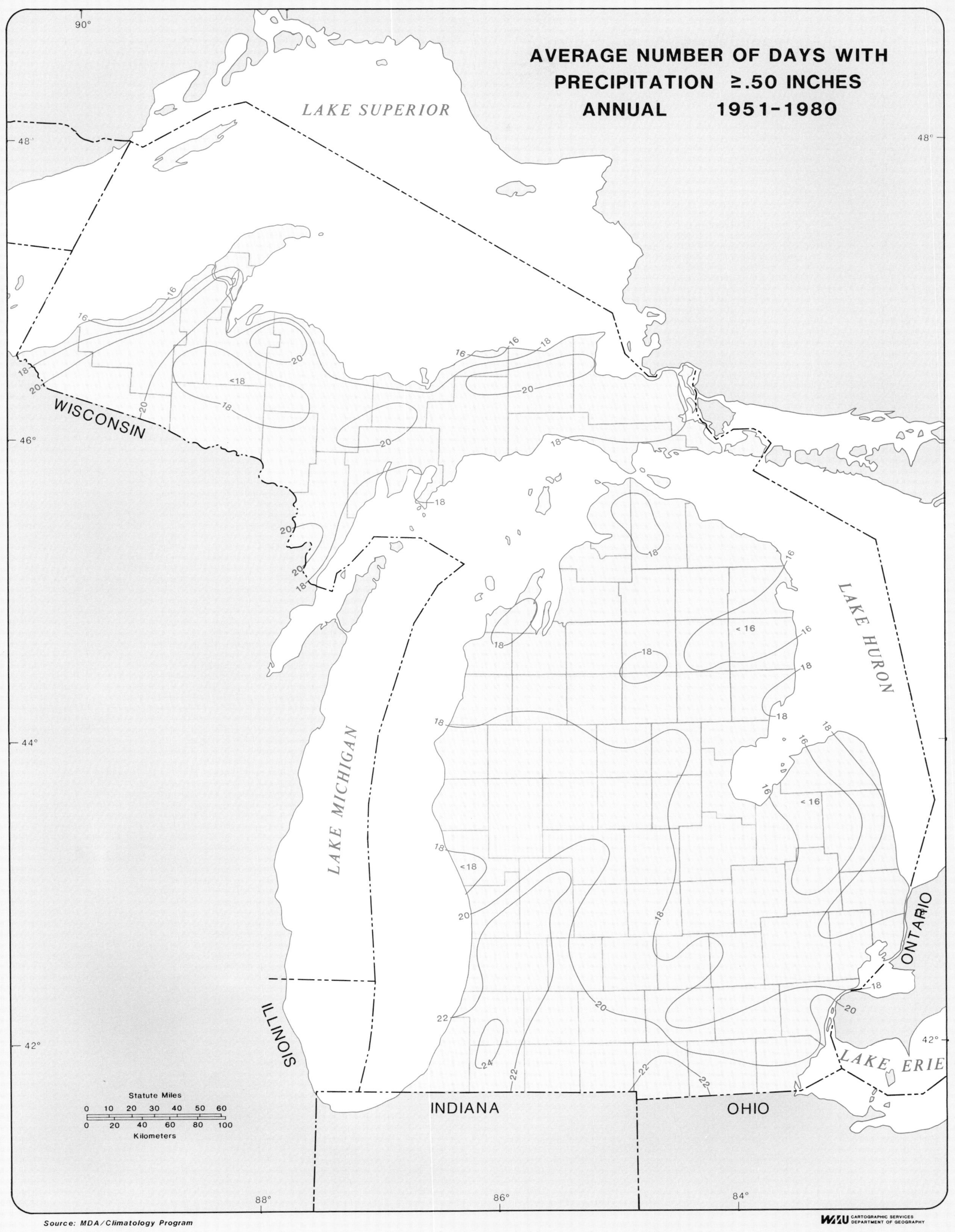


Source: MDA/Climatology Program

WMU CARTOGRAPHIC SERVICES DEPARTMENT OF GEOGRAPHY

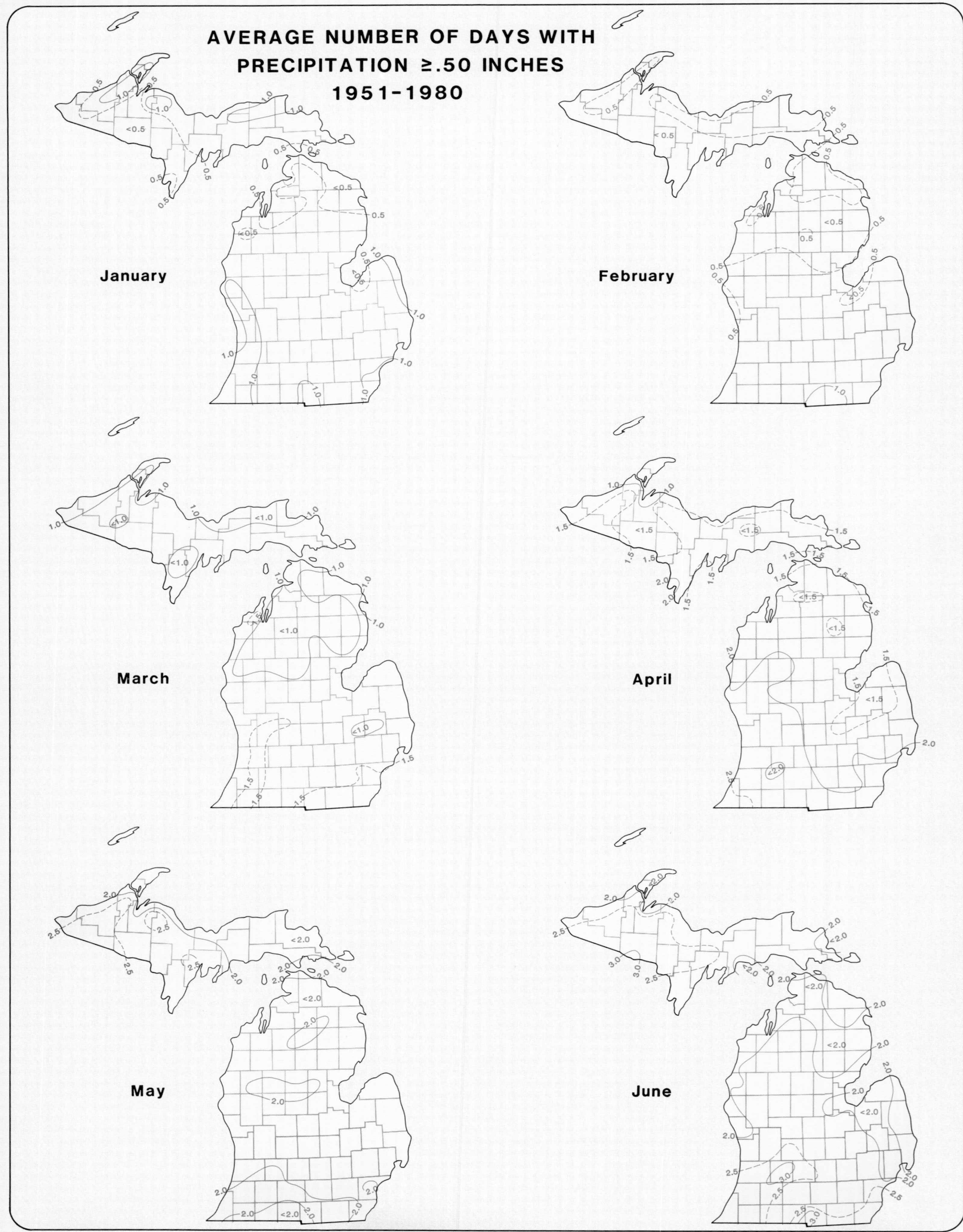


Source: MDA/Climatology Program

WMU CARTOGRAPHIC SERVICES DEPARTMENT OF GEOGRAPHY

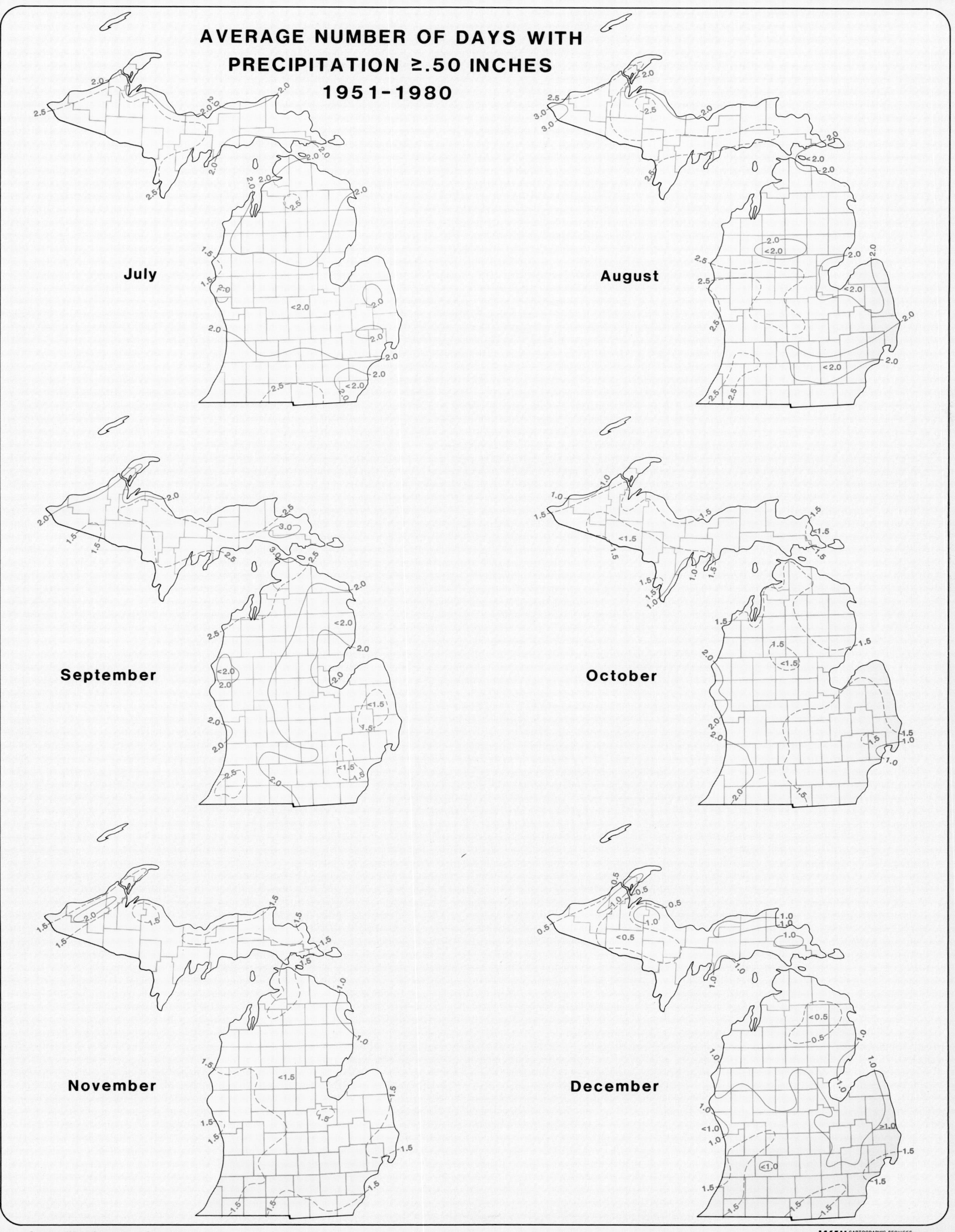


Source: MDA/Climatology Program

WMU CARTOGRAPHIC SERVICES DEPARTMENT OF GEOGRAPHY

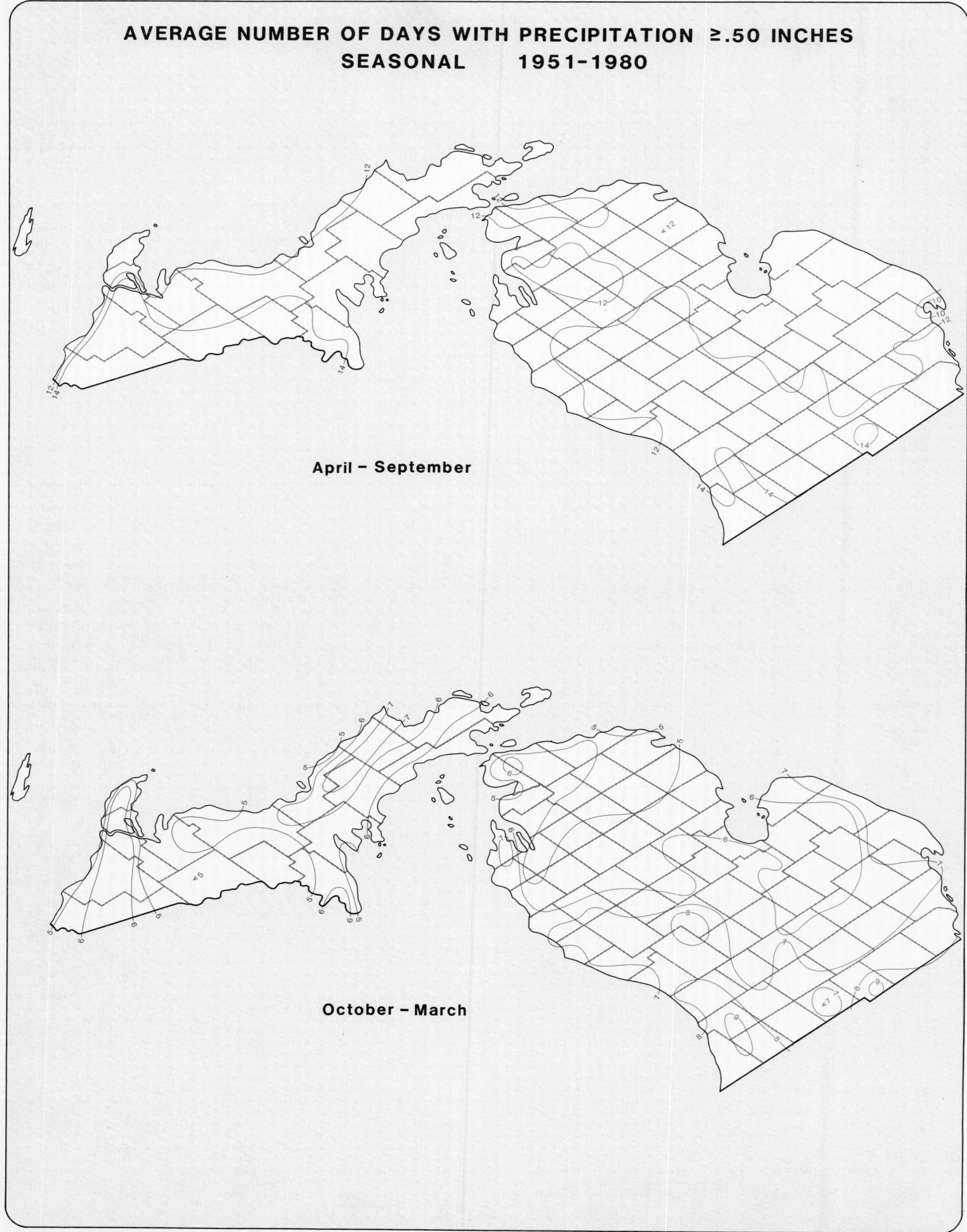


Source: MDA/Climatology Program

WMU CARTOGRAPHIC SERVICES DEPARTMENT OF GEOGRAPHY

# Snowfall

Michigan is known for its snowy winters and some of the heaviest seasonal totals in the eastern United States occur within the state. Many atmospheric disturbances have paths across the Great Lakes causing abundant winter precipitation, and Michigan's northerly latitude insures cold temperatures so that much of the winter precipitation occurs as snow, rather than rain. In addition, the effects of the Great Lakes are of major importance in causing localized snows (lake effect) on downwind shorelines. Lake effect snowfall exerts a major control on the seasonal and monthly snowfall distributional patterns in Michigan and must be considered an important element of the winter weather complex.

Lake effect snow results from modifications of cold air by the warmer open water surfaces of the Great Lakes. The air is warmed by heat transfer from the lakes and becomes more moist because of evaporation from the lake. The warmed and more moist air becomes unstable, rises, and forms clouds which may cause snow over the downwind portions and shorelines of the lakes extending inland 30–40 miles. Convergence caused by land breezes blowing from the colder land to the warmer lake waters may also contribute to the occurrence of lake effect snow, and increased surface friction on downwind shores may cause air to pile up and undergo additional lifting.

Lake effect snowfall may be highly localized as strong storms typically show a banded distributional pattern, causing some locations to receive heavy snow while nearby areas receive little or none at all. The longer the fetch (the distance across water which the air must pass) and the greater the air-water temperature difference, the greater the likelihood of significant lake effect snowfall. If higher land exists on the downwind shore, the air is forced to undergo additional lifting that further intensifies the snowfall.

## *Seasonal*

Because of differences of latitude and lake proximity, average seasonal snowfall totals within the state vary widely. The Detroit area and the extreme southeastern part of the Lower Peninsula average less than 30 inches, while the Keweenaw Peninsula of the Upper Peninsula averages over 200 inches. The largest seasonal snowfall total in Michigan was recorded by a State Highway Snow Station at Delaware, on the Keweenaw Peninsula, which received 391.9 inches during the 1978–1979 season.

The increase in snowfall from southeast to northwest is not an even one as areas of heavier snow (snowbelts) occur along the lake shores in western and northwestern Lower Michigan, and an area in the Upper Peninsula along the north shore of Lake Michigan has lighter snowfall. Thus, Escanaba in the Upper Peninsula has a seasonal average of only 50.3 inches, while Kalamazoo, in southwestern Lower Michigan, averages 73.6 inches.

The snowbelts are sharply defined and occur in both the Lower and Upper Peninsulas. They are areas along the downwind shores of the lakes where lake effect snows are prominent. The snowbelts not only have larger seasonal totals, but have more days with measurable snow and heavier single day totals.

Several snowbelts exist in the Lower Peninsula. To the lee of Lake Michigan in southwest Michigan portions of Muskegon, Ottawa, Allegan, and Van Buren counties have average seasonal amounts over 90 inches, and, in some locations, over 100 inches. A snowbelt in portions of Benzie and Manistee counties extending northward through the Leelanau Peninsula has seasonal totals over 100 inches, and, in some areas, over 150 inches. The elevated areas to the lee of Grand Traverse Bay, including the Mancelona-Gaylord areas and portions of Antrim, Otsego, and Charlevoix counties, average over 140 inches. A small snowbelt also exists along the Lake Huron shore near Harbor Beach, where totals exceed 80 inches.

Snowbelts also occur all along the south shore of Lake Superior in Michigan's Upper Peninsula. The Keweenaw Peninsula and the elevated area north of Michigamme in Baraga and Marquette counties receive average seasonal totals of over 200 inches. Portions of the eastern Upper Peninsula eastward from Munising to Paradise also average from 160 to 180 inches seasonally.

These snowbelts are most prominent during early winter when lake effect snows are more intense. In late winter, the temperature contrasts between the lakes and surrounding land lessen and lake effect snowfall diminishes. Thus, the maps of the 50th percentile of monthly snowfall show the snowbelts very sharply defined during early and midwinter months, but less so in late winter.

Aftermath of big snowstorm of 1978, Kalamazoo.

### *Monthly*

September is the first month when measurable snow falls in the northernmost areas of Michigan. The percentage of years with snowfall is relatively low. During the 30 seasons analyzed here, 1950/51–1979/80, Sault Ste. Marie had the most occurrences, 4, with the maximum amount of 2.7 inches in 1956. Marquette was second with 3 occurrences and the most September snowfall, 5.1 inches, in 1974.

October, also not shown by the maps, ushers in the snow season with portions of the higher elevation areas of the Upper Peninsula having 50th percentile values of more than an inch. The Herman area has more than 2 inches.

November is the first month of the snow season when the entire state has a 50% probability of measurable snowfall. The amounts during the 1951–1980 period range from 28.2 inches at Bergland Dam to 2.0 inches or less in the Detroit area and along Lake Erie's western shoreline in the extreme southeastern Lower Peninsula. Based on a shorter period of record at Herman, it is estimated that the higher elevations in that area have a 50th percentile value of over 30 inches. The significant influence of Lake Superior is readily apparent as the values drop to less than 10 inches near the center of the Upper Peninsula and to less than 5 inches near the Lake Michigan shore.

December ushers in the heart of the snow season with lake effect activity strongly in evidence. The snowbelt areas of the Upper Peninsula have 50th percentile values of over 40 inches while those along the Lake Michigan shoreline in the Lower Peninsula are in the 20 to 30 inch range. The remainder of the Upper Peninsula has less than 20 inches and the Lower Peninsula has less than 15 inches.

January is the month of heaviest snowfall. The higher elevations in the western Upper Peninsula have 50th percentile values of over 60 inches while other lake effect snow areas range from 40 to 60 inches. The remainder of the Upper Peninsula has less than 20 inches. In the Lower Peninsula, the heaviest snow area is in the Leelanau Peninsula with values of 40 inches and more. The other snowbelt areas have more than 30 inches. The rest of the non–lake effect areas of the Lower Peninsula have 15 inches or less and the extreme southeastern part of the state has only about 6 inches.

The downturn in monthly snowfall starts with February. The temperature difference between the water and overpassing air reaches a minimum and ice cover is at a maximum. The fact that February is a short month has some secondary effect. Most heavy snow areas have 30 to 40% less snowfall in February than in January. The February values in the Upper Peninsula range from more than 30 inches to less than 10 inches in the southernmost sections. In the Lower Peninsula, the heavy snow areas are in the 20 to 25 inch range with more than 15 inches along the central Lake Michigan shoreline and "thumb" shoreline of Lake Huron. Most of the interior areas of the Lower Peninsula have 10 inches or less with the extreme southeastern area having less than 6 inches.

March totals continue the downward spiral into spring, especially in the Lower Peninsula where 50th percentile amounts even in the heavier snow areas are only slightly over 15 inches. The major portion of the remainder of the Lower Peninsula has under 10 inches. The Saginaw Bay area and extreme southeastern areas have less than 5 inches. The Upper Peninsula still has considerable snowfall with the heavier snow areas receiving 20 inches or more. More than half of the Upper Peninsula has 15 inches or less while the interior areas have 10 inches or less.

April is the last month of the snow season. The entire state had measurable snowfall during the 30-year period and the 50th percentile values are greater than 1 inch in almost all of the Upper Peninsula and over three-quarters of the Lower Peninsula. The heavier snow areas of the Upper Peninsula have 50th percentile values of greater than 5 inches. In the Lower Peninsula only the highland areas of the north central portion exceed 5 inches.

May, not shown, does not in general have 50th percentile measurable snowfall amounts. Of the 136 stations analyzed, only Houghton FAA has such a value, .2 inch. It is estimated that the higher elevations near and northeast of Herman would have similar amounts. Most of the higher elevations in the western Upper Peninsula averaged 4 to 5 years out of 10 with measurable snowfall in May, while the snowbelt areas of the eastern Upper Peninsula averaged 2 to 3 years in 10. In the Lower Peninsula the higher elevations of the northern sections averaged 1 to 3 years in 10. South of a line from Manistee to East Tawas, only 26 of 73 stations had measurable snowfall in May, occurring on only one or two years during the 30-year period.

Snowfall has occurred as late as June in rare instances at some of the higher elevations but melts almost as soon as it hits the warm ground. Other reports of snowfall during the summer months are usually soft hail or snow pellets associated with thunderstorms and are not really considered snow.

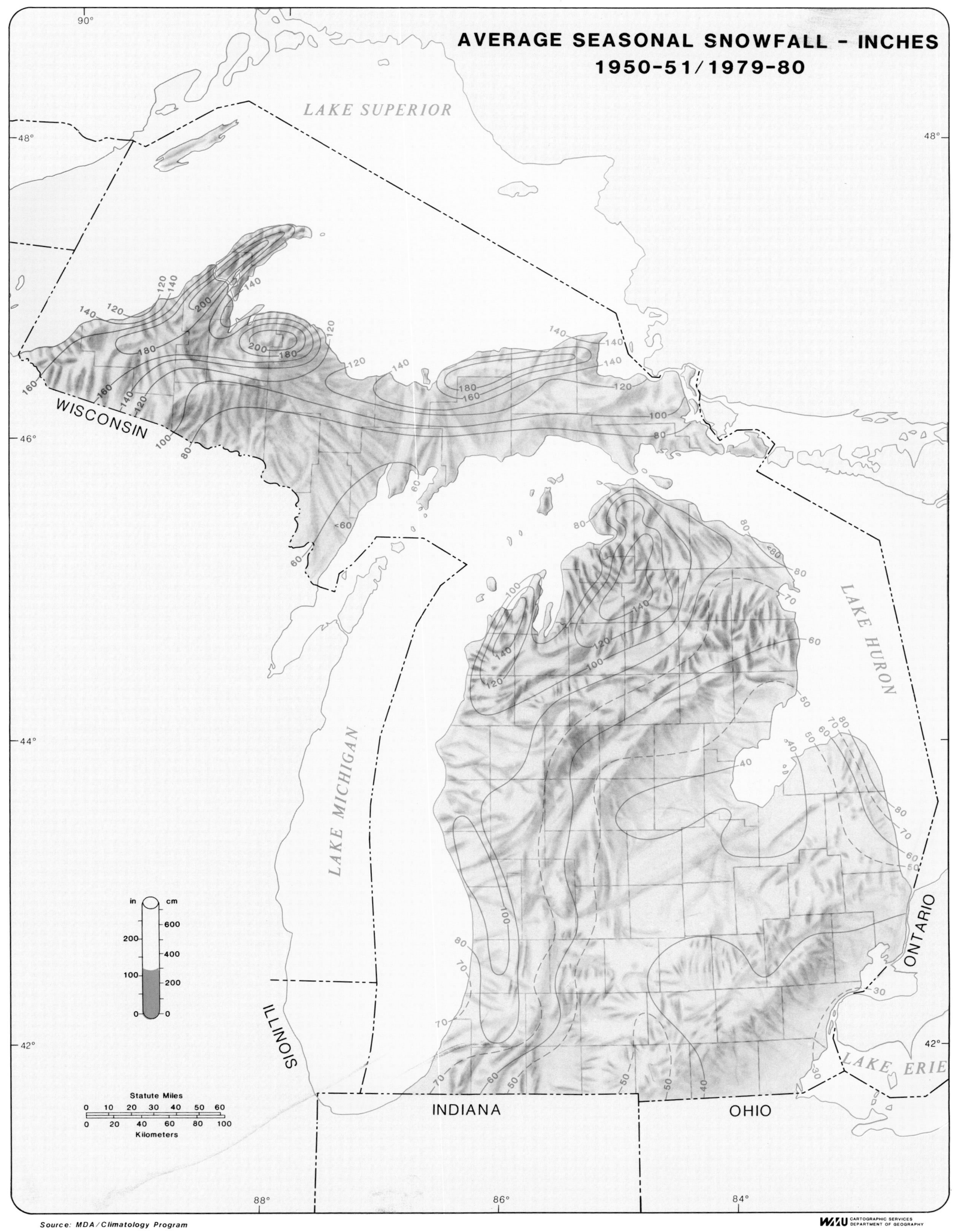
AVERAGE SEASONAL SNOWFALL - INCHES
1950-51/1979-80
LAKE SUPERIOR
WISCONSIN
LAKE MICHIGAN
LAKE HURON
ILLINOIS
INDIANA
OHIO
ONTARIO
LAKE ERIE
Statute Miles
Kilometers
Source: MDA/Climatology Program
CARTOGRAPHIC SERVICES DEPARTMENT OF GEOGRAPHY

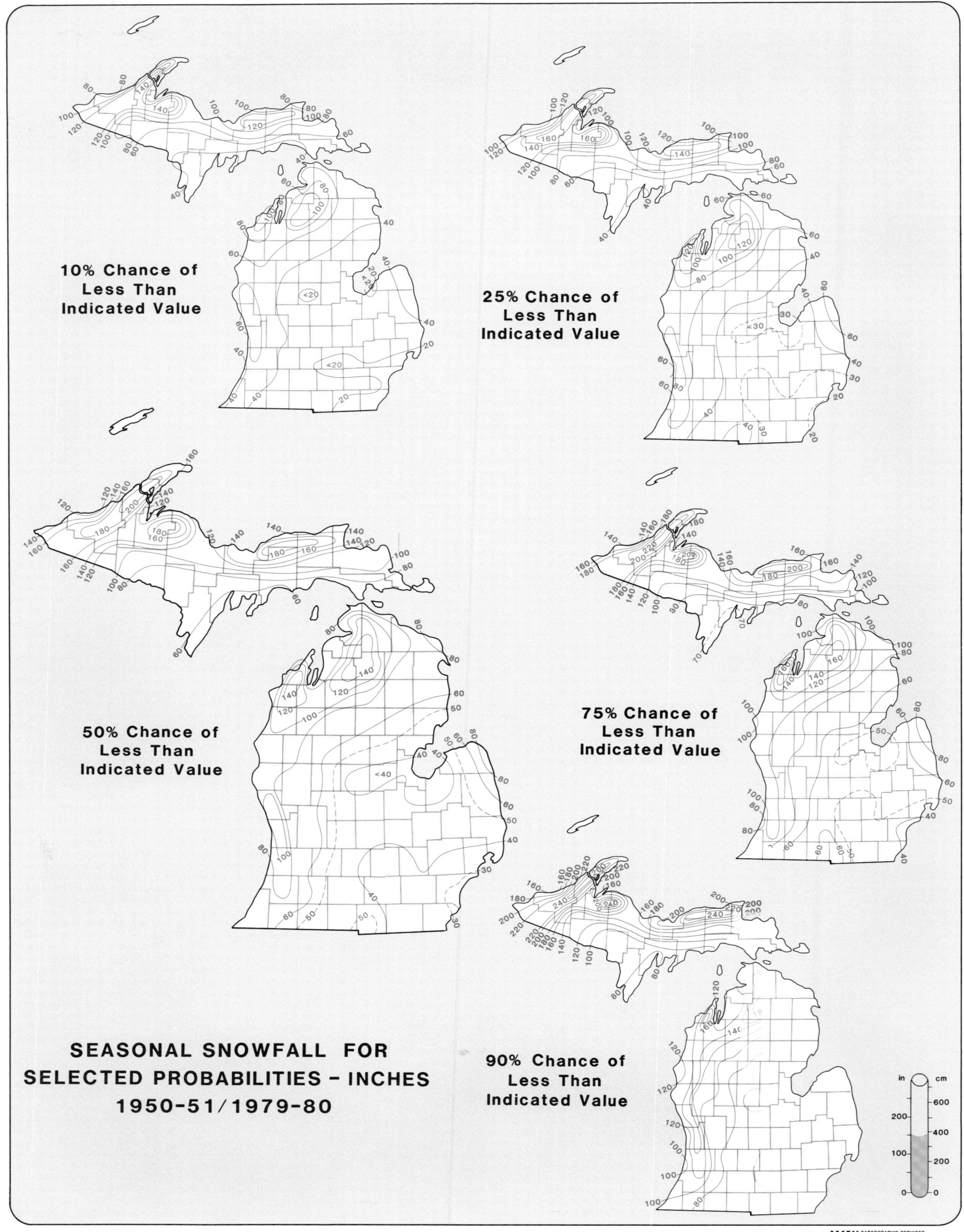


Source: MDA/Climatology Program

WMU CARTOGRAPHIC SERVICES DEPARTMENT OF GEOGRAPHY

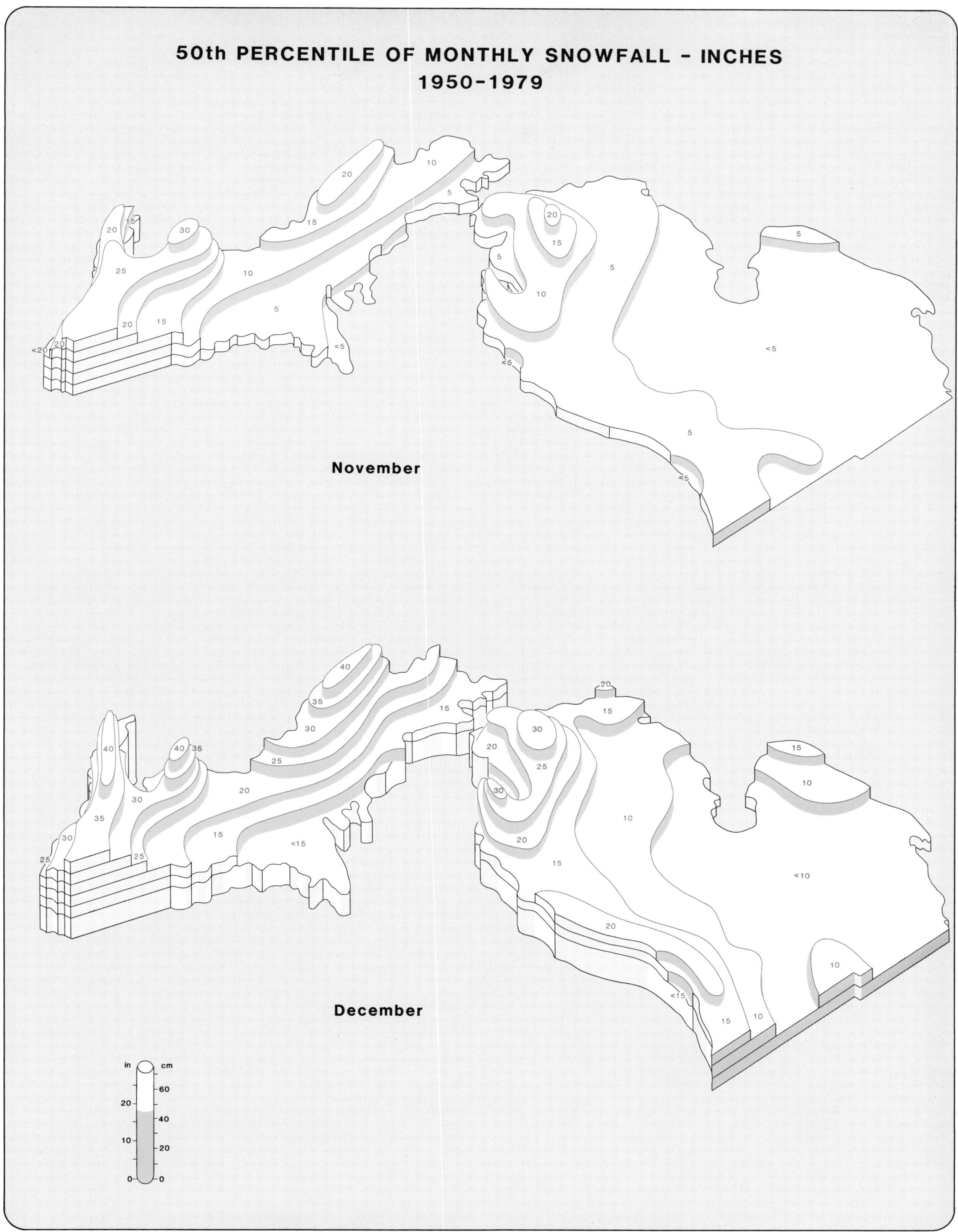


Source: MDA/Climatology Program

WMU CARTOGRAPHIC SERVICES DEPARTMENT OF GEOGRAPHY

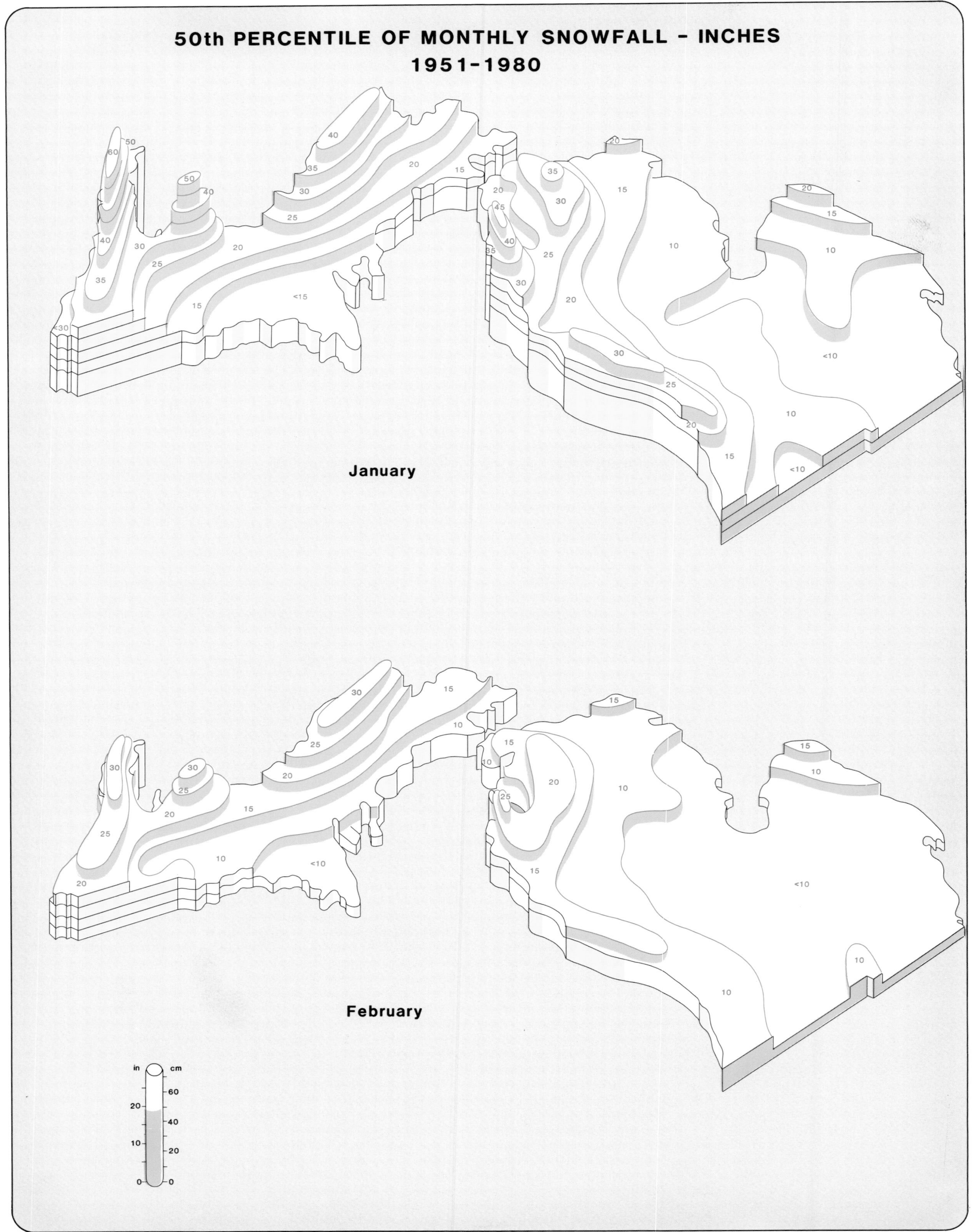


Source: MDA/Climatology Program

WMU CARTOGRAPHIC SERVICES DEPARTMENT OF GEOGRAPHY

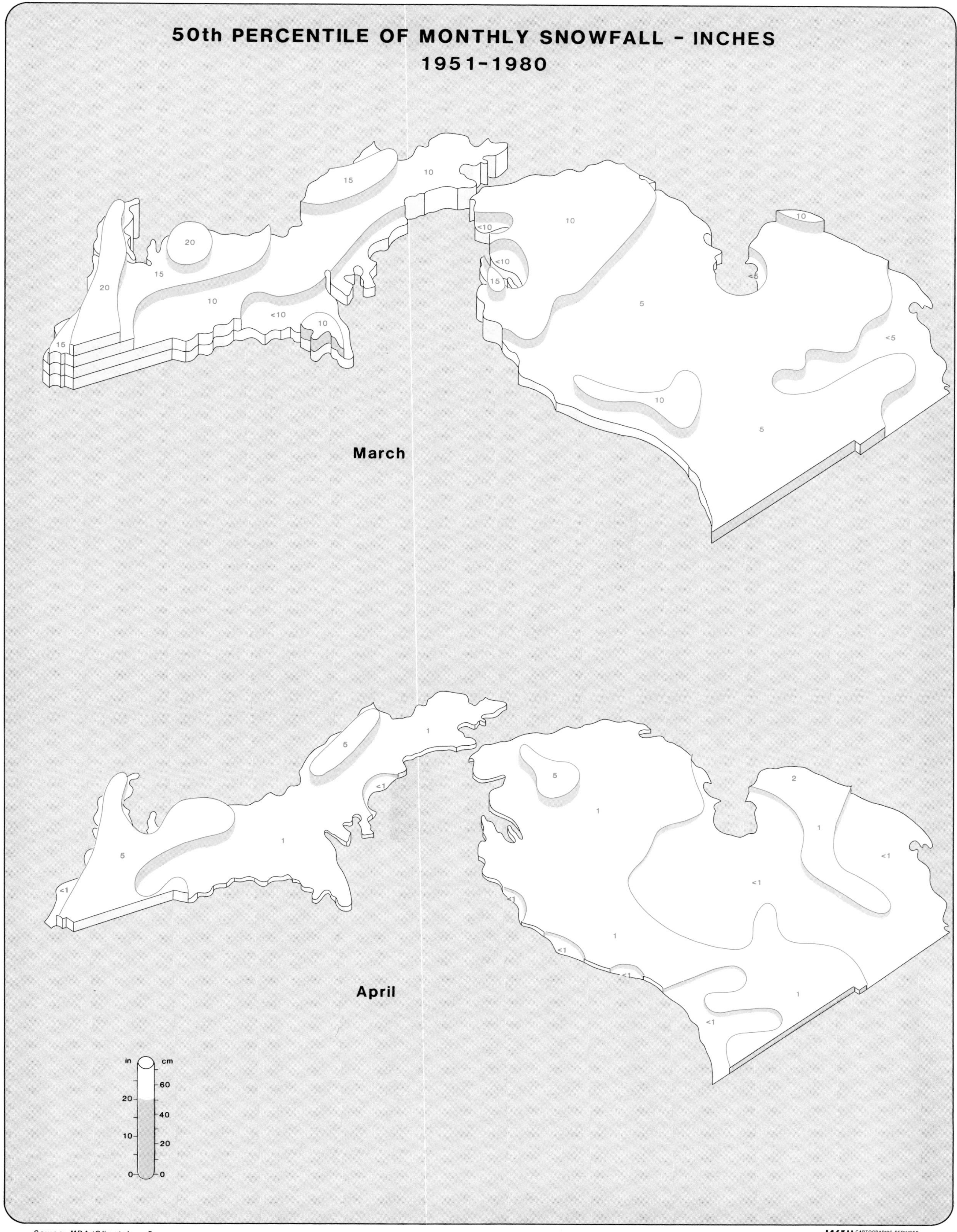


Source: MDA/Climatology Program

WMU CARTOGRAPHIC SERVICES DEPARTMENT OF GEOGRAPHY

Lightning.

# THUNDERSTORMS

Thunderstorms require warm, moist air for their development. Consequently, they occur infrequently in Michigan during the colder months, but are rather common during the warm season. They generate both costs and benefits for the state. Much of the warm season rainfall occurs as a result of thunderstorms, providing invaluable moisture for agriculture and often spelling relief from heat and humidity. On the other hand, severe thunderstorms may spawn high winds, damaging hail, and tornadoes. Cloud to ground lightning strikes accompanying thunderstorms are counted among the leading weather-related causes of fatalities.

The maps showing average annual and monthly mean frequencies of thunderstorms (thunder events) in Michigan were redrawn from national maps appearing in the publication *National Thunderstorm Frequencies for the Contiguous United States* by M. J. Changery. The data were extracted from original manuscript sources at stations taking hourly weather observations including National Weather Service, FAA, Air Force, Navy, and Marine Corps stations. Michigan stations included Detroit City Airport, Flint, Grand Rapids, Houghton, Jackson, Lansing, Muskegon, Pellston, Sault Ste. Marie, Traverse City, and Wurtsmith AFB near Oscoda. A 30-year period, 1948–1977 was used as a base, although this varied slightly with some stations.

The maps show the strong seasonal contrast in thunderstorm occurrences in Michigan. December and January have the least thunderstorms and June and July the most. The southern half of the Lower Peninsula has increased thunderstorm occurrence as temperatures warm in March and April, while thunderstorms still occur infrequently in the Upper Peninsula and along the south shore of Lake Superior. Also noticeable during the spring months is the role of the lakes in restricting thunderstorm occurrence along their immediate shores.

Building thunderstorm.

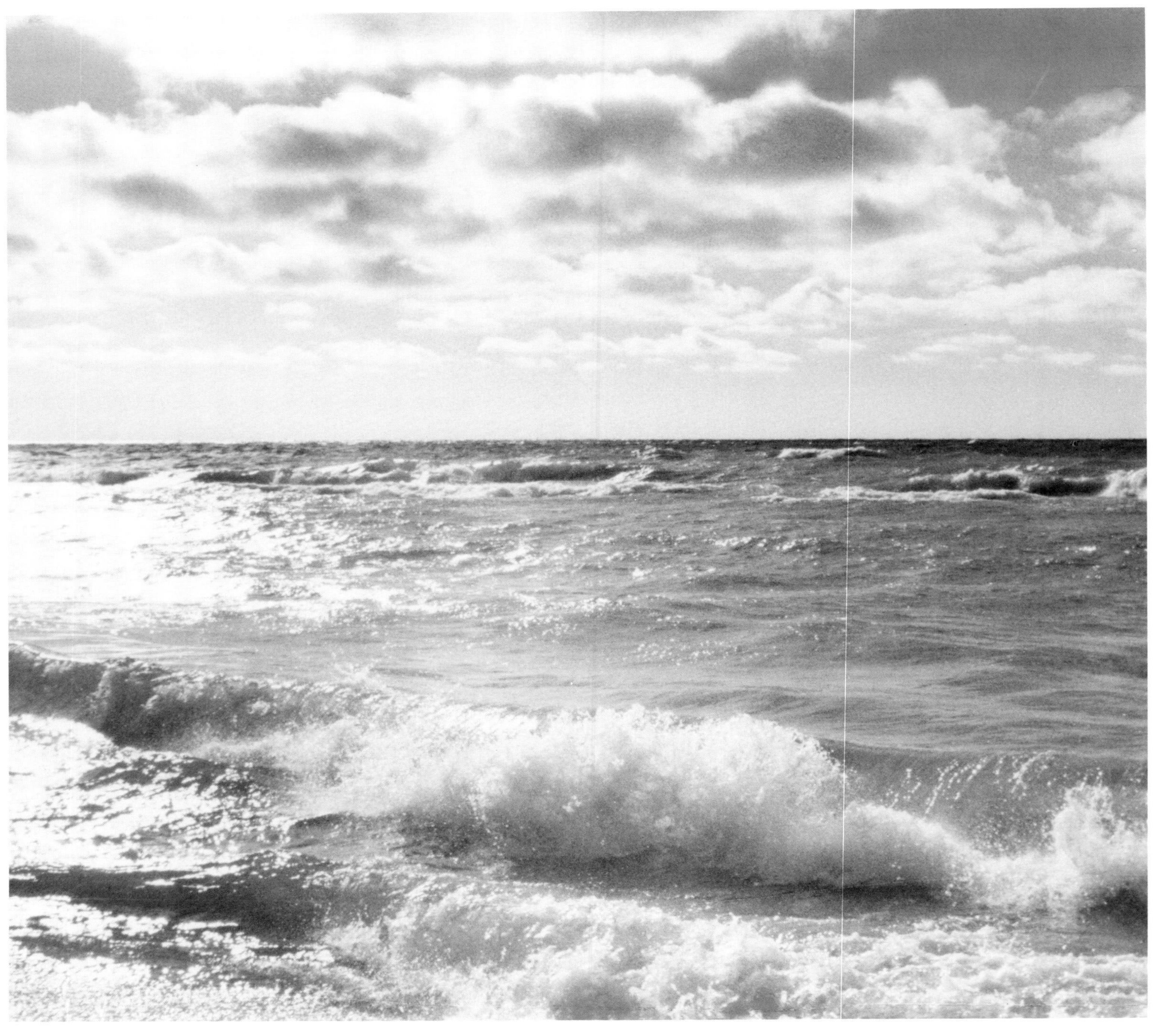

Lake Michigan.

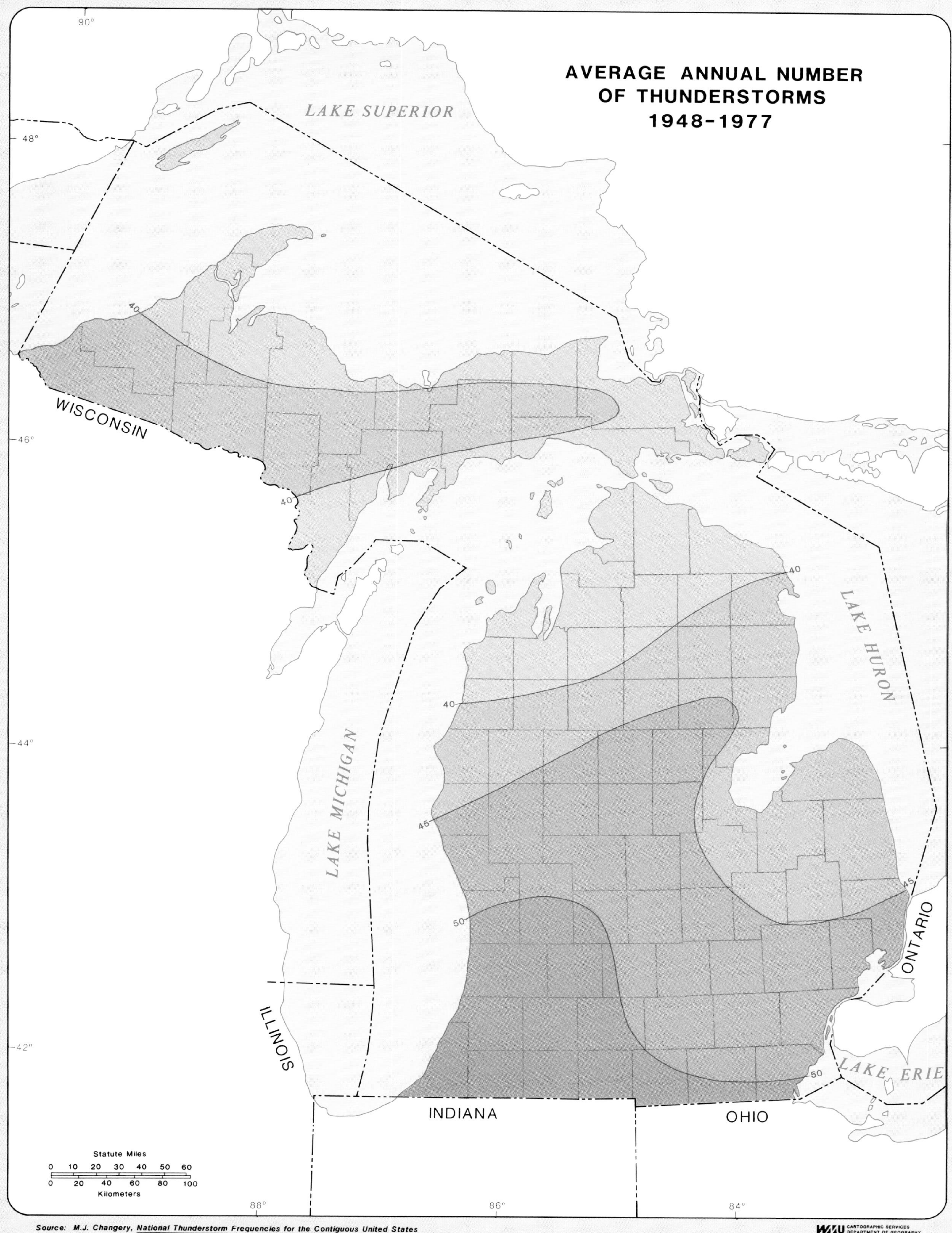


Source: M.J. Changery, National Thunderstorm Frequencies for the Contiguous United States

WMU CARTOGRAPHIC SERVICES DEPARTMENT OF GEOGRAPHY

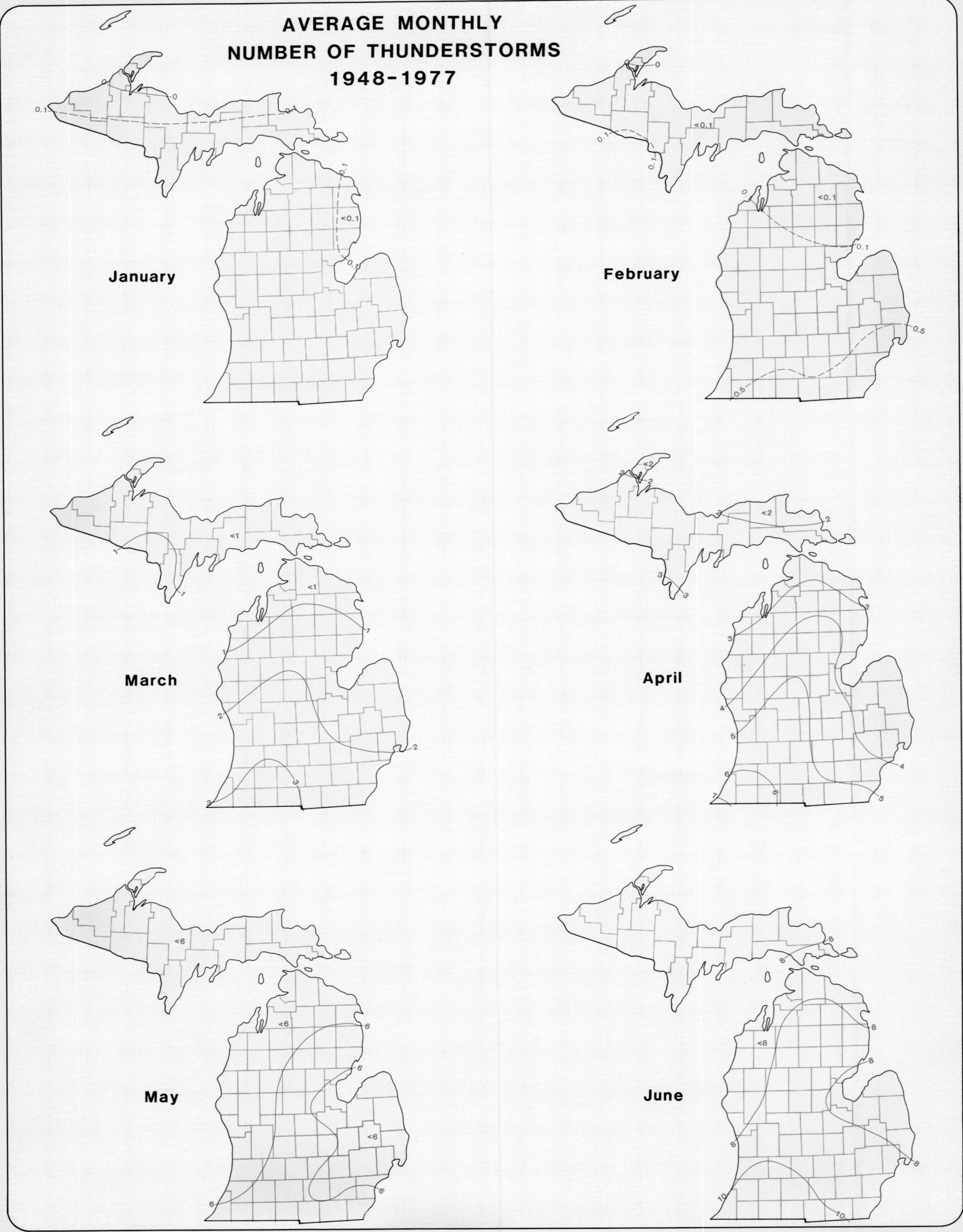


Source: M.J. Changery, National Thunderstorm Frequencies for the Contiguous United States

WMU CARTOGRAPHIC SERVICES DEPARTMENT OF GEOGRAPHY

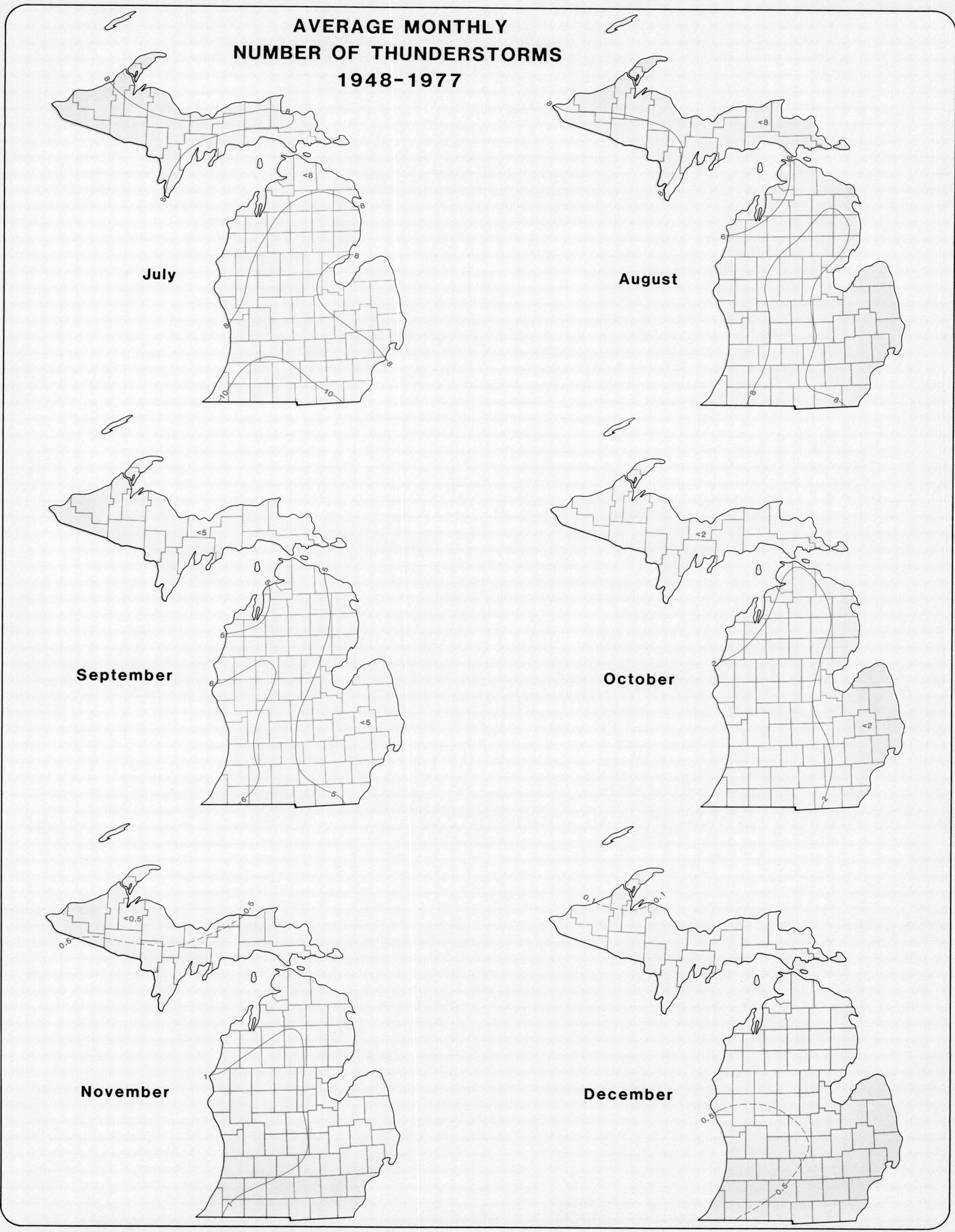


Source: M.J. Changery, National Thunderstorm Frequencies for the Contiguous United States

WMU CARTOGRAPHIC SERVICES DEPARTMENT OF GEOGRAPHY

Mammatus sky: indication of severe weather.

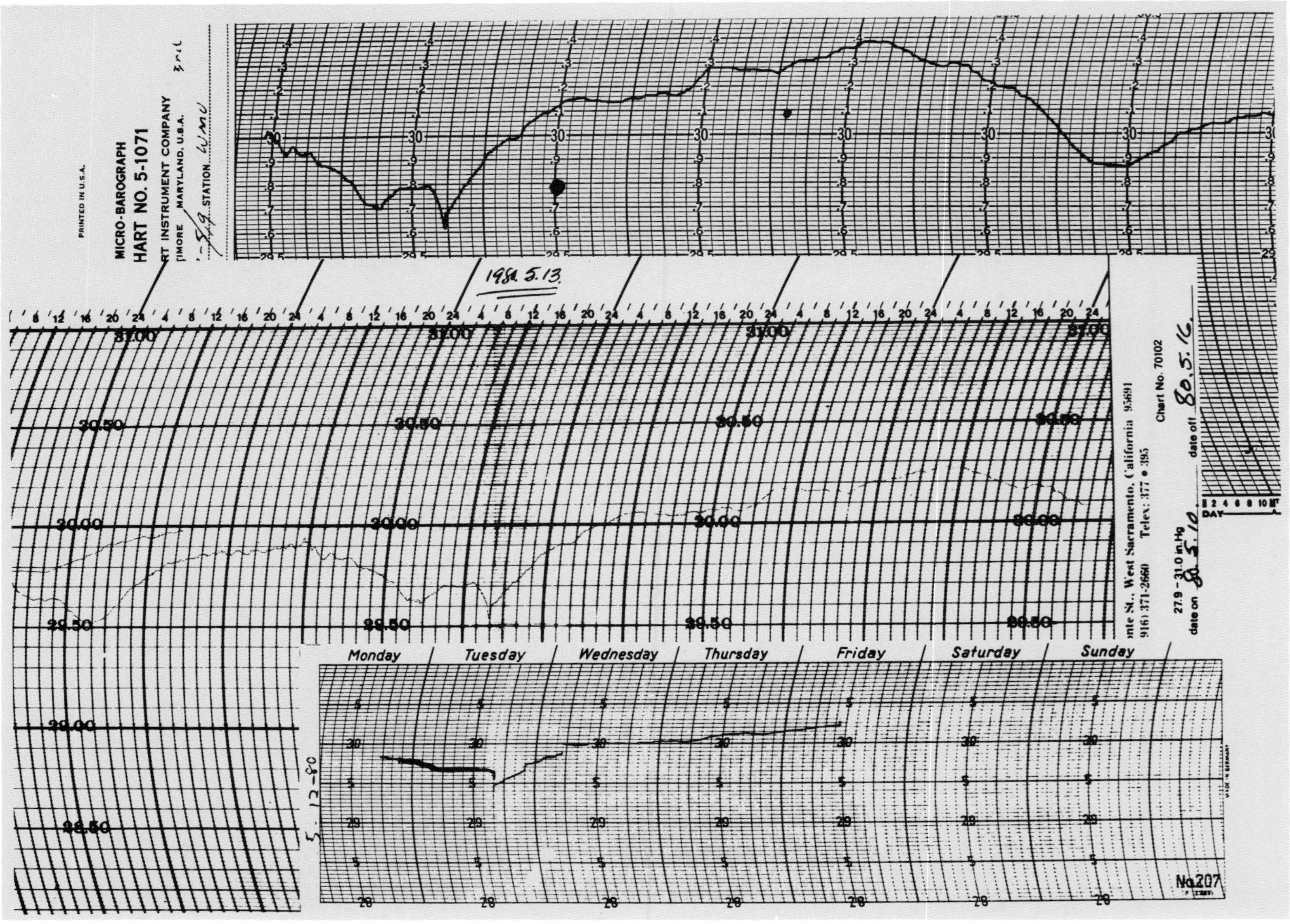

Barograph recordings of Kalamazoo tornado on May 13, 1980, at various distances from path of destruction. Upper 1 km, middle 2.5 km, lower 0 km.

# TORNADOES

## Tornado Touchdowns and Paths

Although tornadoes can occur at any place in the state, records of the 51 years, 1930–1980, show that tornadoes have touched down most often in the southern Lower Peninsula, less in the northern Lower Peninsula, and least in the Upper Peninsula. Moreover, a six-by-six mile recording network (shown on the probability map) indicates that numbers of tornado hits vary greatly within these regions. For example, some grid cells in the southern region were struck by as many as 6 tornadoes during 1930–1980, while other cells were hit fewer times, or not at all. The following formula was utilized to calculate probabilities of tornado recurrence:

$$P = \frac{Ea * t}{A}$$

Ea = mean path area
t = cell's mean annual number of tornadoes
A = cell's area in square miles.

For cells with one tornado during 1930–1980, the calculated probability of 0.0011 represents a recurrence ratio of 918 years before a tornado will strike a given point inside the 36 square mile area again. A point inside a cell having six tornadoes, on the other hand, might be hit an average of once in 153 years.

Funnel cloud.

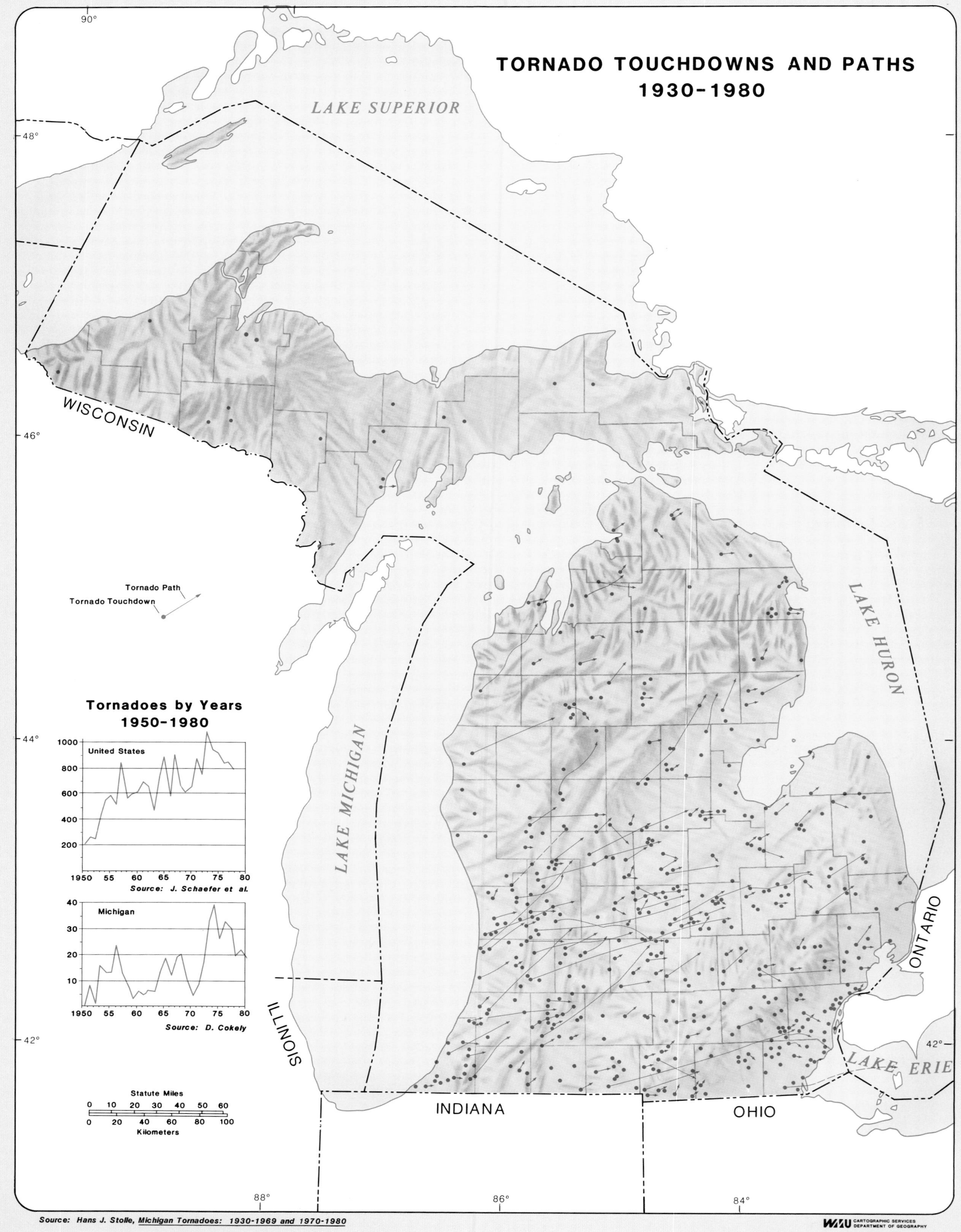


Source: Hans J. Stolle, Michigan Tornadoes: 1930-1969 and 1970-1980

WMU CARTOGRAPHIC SERVICES DEPARTMENT OF GEOGRAPHY

# Number of Tornadoes, Tornado Days, Fatal Tornadoes and Tornado Deaths by Month, 1930–1969, 1970–1980

When warm and moist air collides with cold and dry air, tornadoes can occur. In Michigan, tornadoes did not occur in December and January and only rarely in November and February during the 51-year study period. This was due to the scarcity of warm air during these months.

During a typical year, the number of tornadoes tends to increase quickly in spring and drop off rapidly in the fall. However, a comparison of monthly numbers of Michigan tornadoes during 1930–1969 and 1970–1980 points out a change in the monthly frequencies of tornadoes from a skewed distribution peaking in the spring to a more normal distribution peaking in early summer.

During the earlier period, a small number of tornadoes in March was followed by an eighteenfold increase to a peak in April. In May, this number dipped slightly but rebounded to peak again in June. A sudden decrease in July by more than one half was followed by a 30% increase in August and a steady decline during September and October.

During the 1970–1980 period, on the other hand, there were almost five times as many tornadoes in March. Tornado occurrences then increased by 70% in April, but there were far fewer April tornadoes than during the earlier period. After a slight decrease in May, tornadoes more than doubled to peak in June and then declined steadily during July, August, September, and October.

In contrast, the two periods' distributions of monthly numbers of tornado days are almost normal, and both periods' frequencies of fatal tornadoes are skewed with peaks in spring.

# Diurnal Frequencies of Tornadoes, 1930–1969, 1970–1980

Tornadoes can occur at any hour. Nonetheless, a tabulation of all 51-year tornado occurrences by time of day produced a near normal distribution, peaking in late afternoon. When comparing the daily distributions of 1930–1969 with those of 1970–1980, a four-hour shift in peak time can be observed. In search for an explanation, all tornado time-day data were sorted and plotted by months. The results revealed that spring and fall tornadoes tend to occur later in the day than summer tornadoes do. And, since more 1970–1980 tornadoes occurred in the summer than did in the 1930–1969 period, an earlier overall daily time distribution resulted.

# Tornado Path Directions and Lengths

Plotting the directions of more than 300 tornadoes on a windrose diagram identified several interesting traits of Michigan tornadoes. They tend to move in an east-northeasterly direction and the lengths of their paths decrease with angular distance from the preferred direction.

# Number of Tornadoes by County Since 1930 and Five-Year Increments

To help identify spatial shifts in tornado touchdown patterns, the mapped distribution of 1930–1969 tornadoes is compared with five-year increment distributions since 1970. Besides a general increase for the entire state, large increase counties are located in the southern part of the Lower Peninsula. Moreover, the counties showing the largest increases are at the Ohio and Indiana borders. Hence a shift within the state of the high frequency tornado impact areas toward the south is evident.

For this section's maps and graphs, pre-1940 tornado data were obtained from the *Monthly Weather Review*, 1930–1950, and *Storm Summaries, Climatological Data, Michigan Section*, 1930–1957. The *Climatological Data, National Summary*, 1950–1958, and monthly issues of *Storm Data*, 1959–1980, were used for later tornadoes.

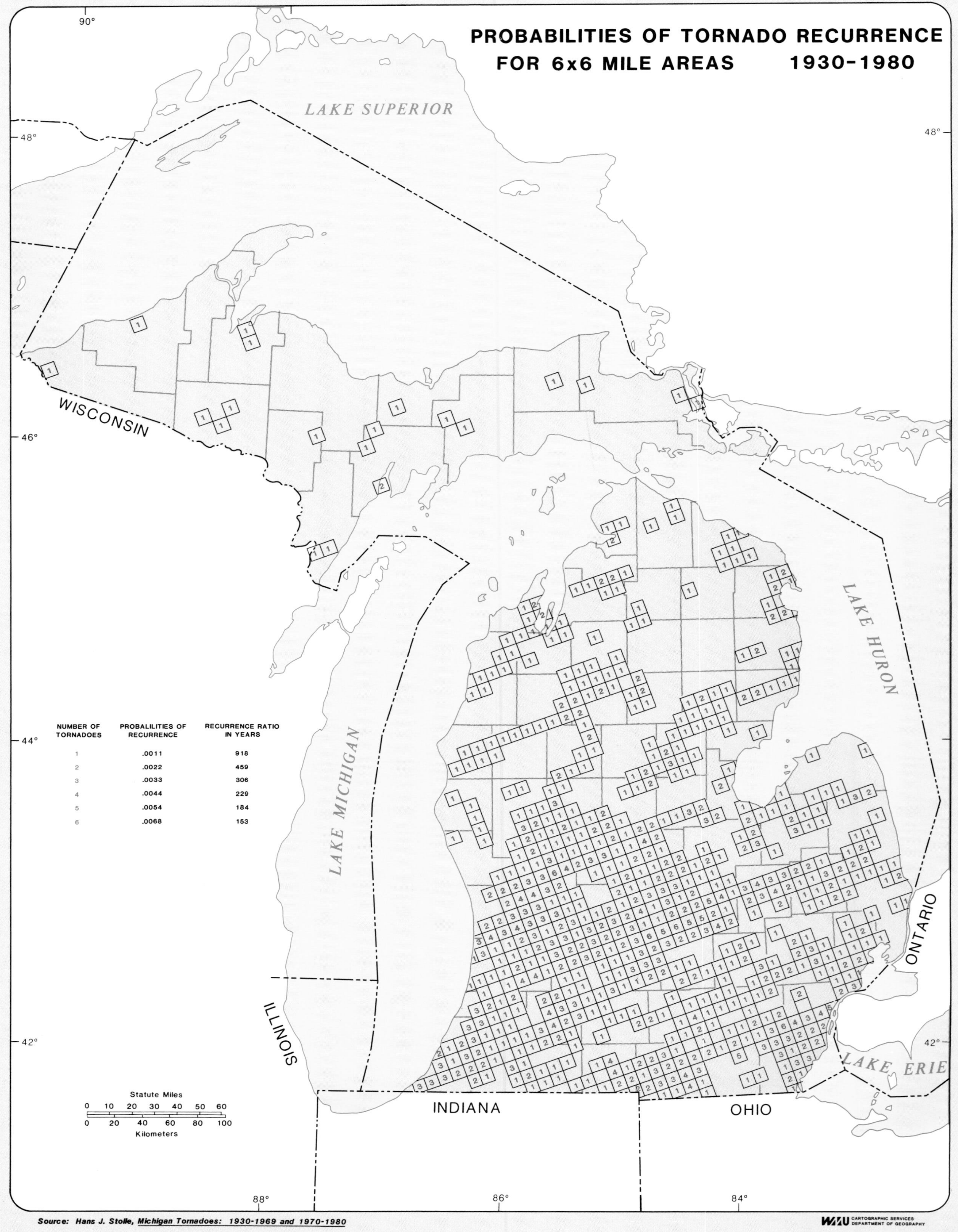


| NUMBER OF TORNADOES | PROBALILITIES OF RECURRENCE | RECURRENCE RATIO IN YEARS |
|---|---|---|
| 1 | .0011 | 918 |
| 2 | .0022 | 459 |
| 3 | .0033 | 306 |
| 4 | .0044 | 229 |
| 5 | .0054 | 184 |
| 6 | .0068 | 153 |

*Source: Hans J. Stolle, Michigan Tornadoes: 1930-1969 and 1970-1980*

WMU CARTOGRAPHIC SERVICES DEPARTMENT OF GEOGRAPHY

## DIURNAL FREQUENCIES OF TORNADOES 1930-69/1970-80

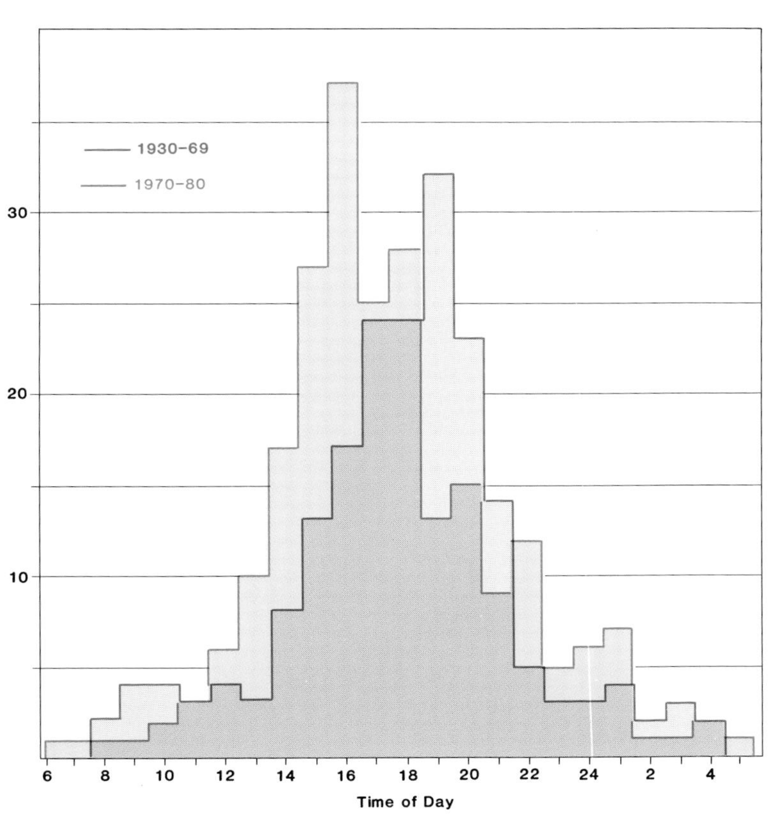


## TORNADO PATH DIRECTIONS AND LENGTHS, 1930-80

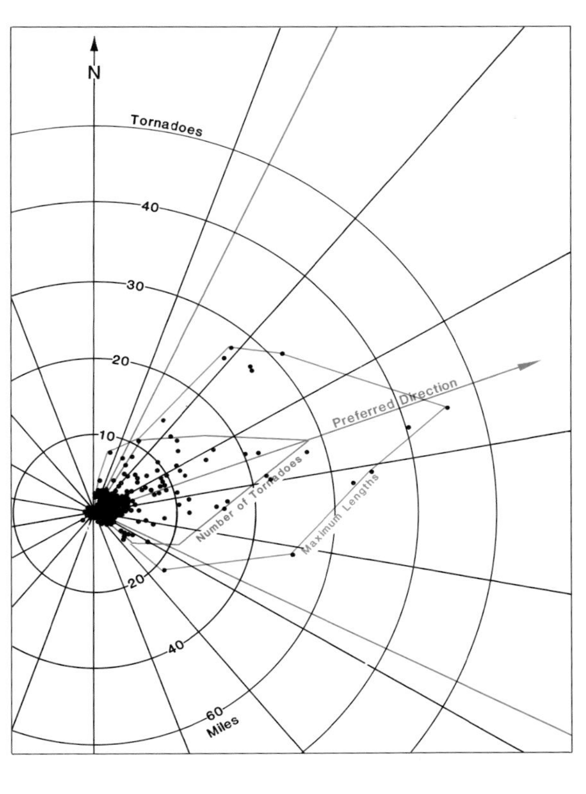


## NUMBER OF TORNADOES, TORNADO DAYS, FATAL TORNADOES AND TORNADO DEATHS BY MONTH, 1930-69/1970-80

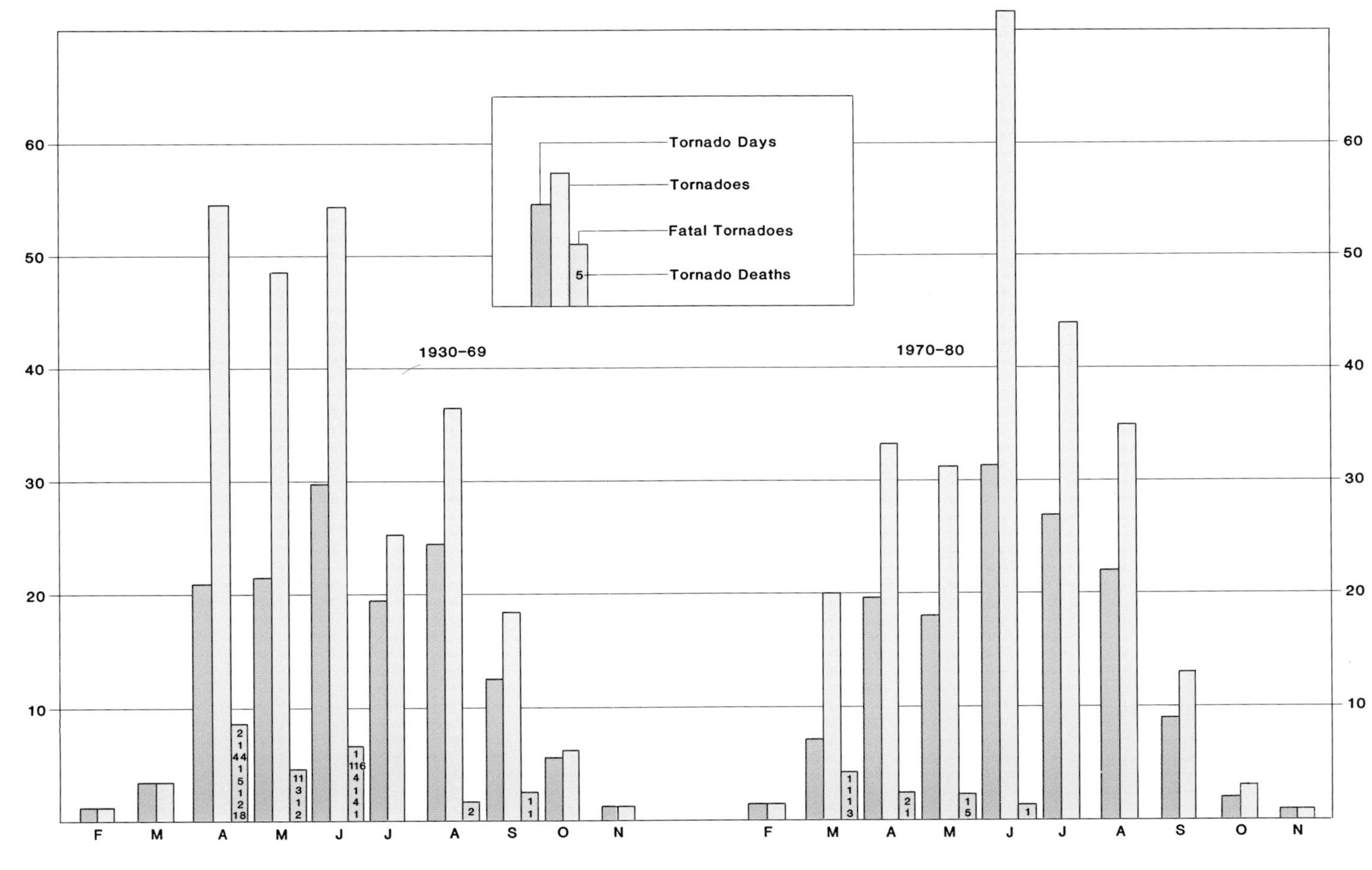


Source: *Hans J. Stolle, Michigan Tornadoes: 1930-1969 and 1970-1980*

WMU CARTOGRAPHIC SERVICES DEPARTMENT OF GEOGRAPHY

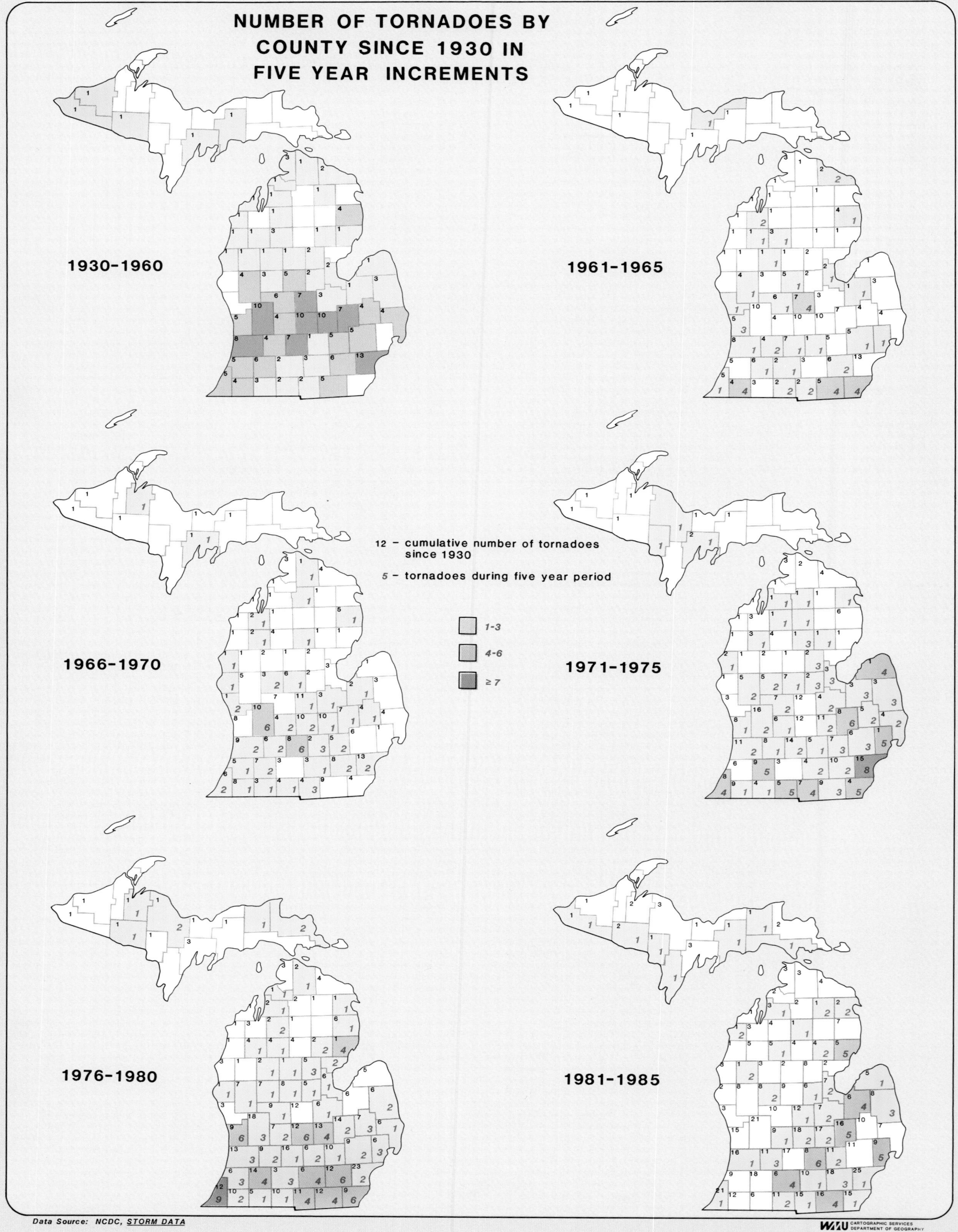
NUMBER OF TORNADOES BY COUNTY SINCE 1930 IN FIVE YEAR INCREMENTS
1930-1960
1961-1965
1966-1970
1971-1975
1976-1980
1981-1985
12 – cumulative number of tornadoes since 1930
5 – tornadoes during five year period
1-3
4-6
≥ 7
Data Source: NCDC, STORM DATA
WMU CARTOGRAPHIC SERVICES DEPARTMENT OF GEOGRAPHY

# EVAPORATION

Limited evaporation data are available for Michigan, although data from Class "A" evaporation pans have been recorded at six stations since 1948 as part of the National Weather Service climatological data network. In general, the ratio of the amount of evaporation from a large lake to that measured by an evaporation pan (pan coefficient) is less than one (typically .75), which means that the evaporation recorded by a pan may overestimate that from a lake or large reservoir. This limitation should be taken into consideration when the data presented here are being interpreted.

The pan evaporation data from the six Michigan stations have been summarized by Fred V. Nurnberger in the publication *Summary of Evaporation in Michigan*. The data used in the construction of the following graphs were taken from that publication. The time period for the stations was variable ranging from 1948–1980 at Seney to 1960–1980 at Lake City. Average monthly precipitation amounts are also shown.

In Michigan, the maximum monthly average pan evaporation occurred in July for all stations, ranging from 7.31 inches at East Lansing to 5.47 inches at Seney. Seasonally, average amounts range from 39.02 inches at East Lansing to 23.43 inches at Seney. In general, evaporation amounts are larger in the south where temperatures are higher, although other factors such as solar radiation, wind, and humidity also play a role in determining evaporation rates.

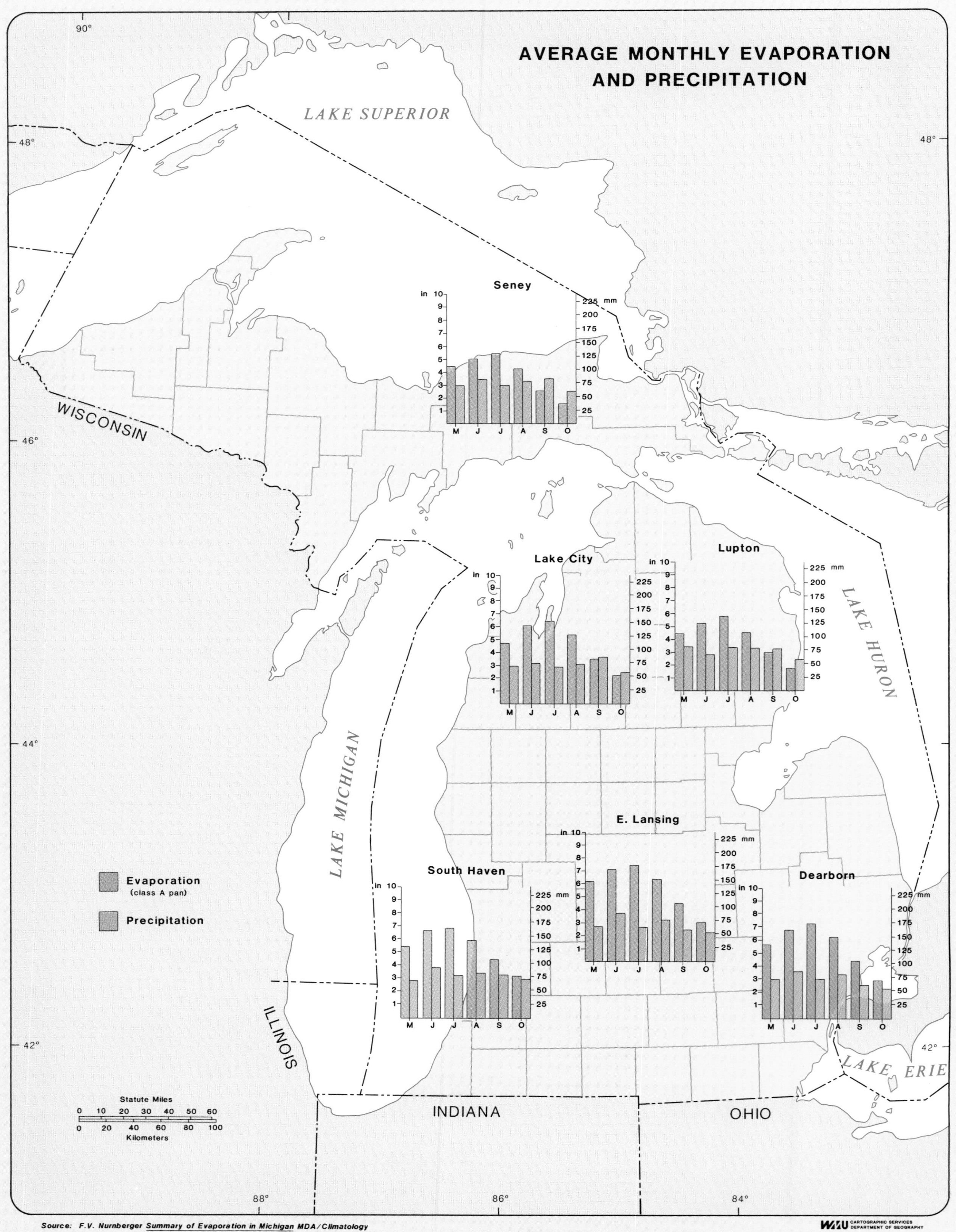


Source: F.V. Nurnberger *Summary of Evaporation in Michigan* MDA/Climatology

WMU CARTOGRAPHIC SERVICES DEPARTMENT OF GEOGRAPHY

# WIND

Wind data for Michigan are available from a number of National Weather Service, airport, and Air Force stations. The Pacific Northwest Laboratory (Battelle Memorial Institute, Richland, Washington) assembled wind data for a number of Michigan stations into a computerized data base that was utilized in the production of a Wind Energy Resource Atlas for the Great Lakes area. The data base for this atlas was archived on microfiche and magnetic tape. The microfiche for seven Michigan stations were obtained from the National Climatic Data Center and the data used in the construction of maps and graphs for the wind section which follows.

The seven stations include Detroit (Metro-Wayne Airport) 1961–1978, Gwinn (K. I. Sawyer AFB) 1958–1970, Lansing (Capital City Airport) 1963–1978, Muskegon (County Airport) 1961–1978, Traverse City (Cherry Cap. Airport) 1962–1978, Sault Ste. Marie (Municipal Airport) 1966–1978, and Houghton (County Airport) 1954–1964.

Cup anemometer .

## Average Annual Wind Directional Frequency and Speed

Michigan lies within the belt of westerlies, and winds prevail from this general direction. During the summer months winds are predominantly from the southwest due to the location over the southeastern United States of the semipermanent Bermuda High. During the winter winds prevail from the west or northwest, although they change frequently for short periods as migrating cyclones and anticyclones move through the area.

There can be considerable variation from station to station, however, depending upon location and exposure. Winds at Sault Ste. Marie and Houghton show a strong component of east and east-southeast winds in addition to large frequencies from the west-northwest. Winds at Gwinn prevail from both the south and north while winds at Traverse City prevail from the south. Wind velocities are highest at most stations with directions between west and north, with the exception of Muskegon, where the highest velocities occur with south and north winds.

## Average Monthly Wind Speed and Wind Power

Wind power (watts $m^2$) was computed by researchers at the Battelle Memorial Institute and is a measure of wind as a power source. It is based on the distribution of wind speeds and the density of air in $Kg/m^3$, computed from station temperature and pressure. Winter is the season of greatest wind speed and power, while summer has the least wind speed and power, with August normally the month of lowest wind speed. In general, the more southerly stations show a greater seasonal variation in wind speed and power. Gwinn and Houghton, in the Upper Peninsula, show spring maxima of wind speed and a tendency towards a secondary maximum in spring occurs at the other stations. The average annual wind speed at anemometer height is greatest at Muskegon, along the Lake Michigan shoreline, with 4.9 meters per second (10.9 mph). Average annual wind speeds are least at Gwinn, with 3.8 mps (8.5 mph).

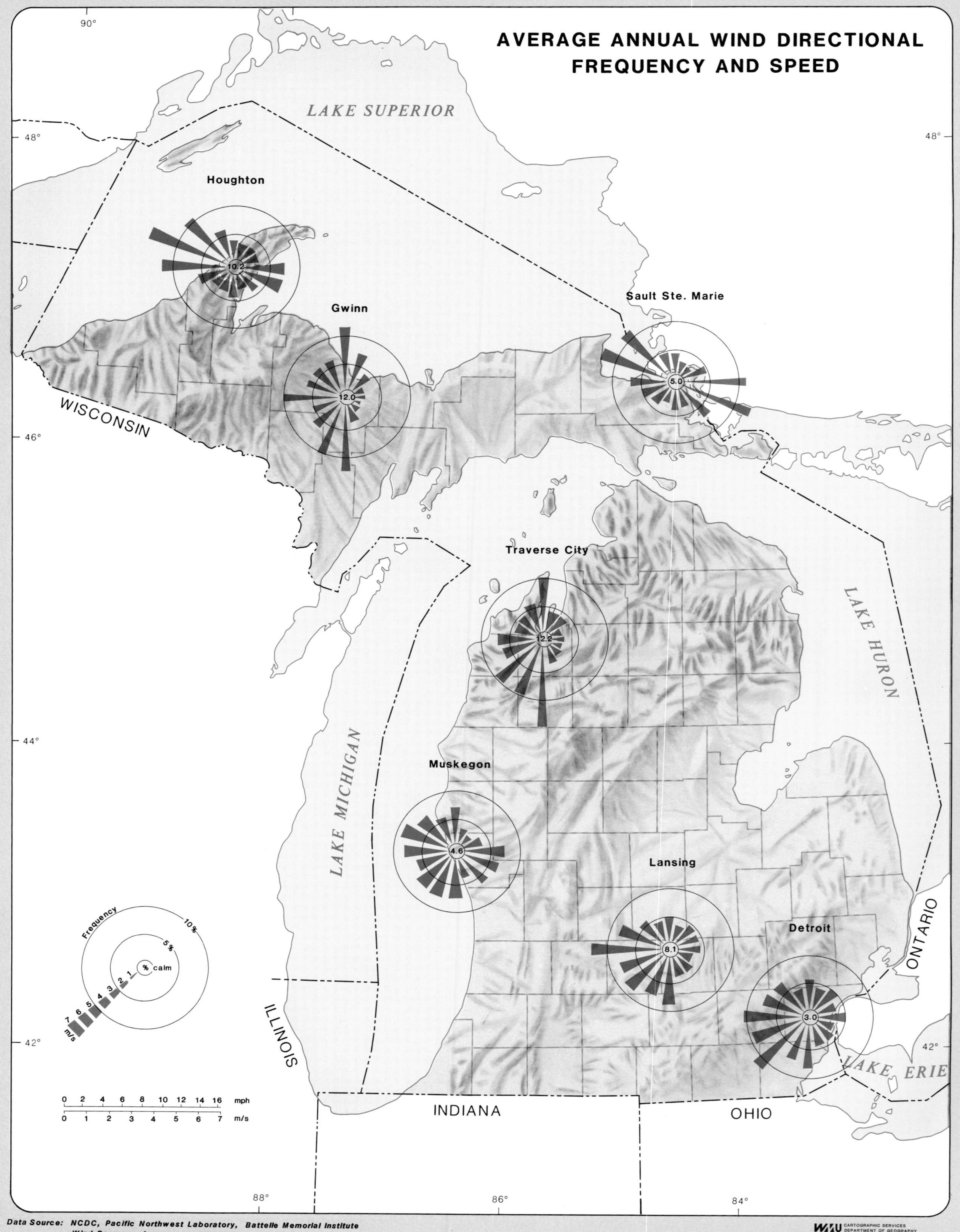


Data Source: NCDC, Pacific Northwest Laboratory, Battelle Memorial Institute Wind Resource Assessment Data File

WMU CARTOGRAPHIC SERVICES DEPARTMENT OF GEOGRAPHY

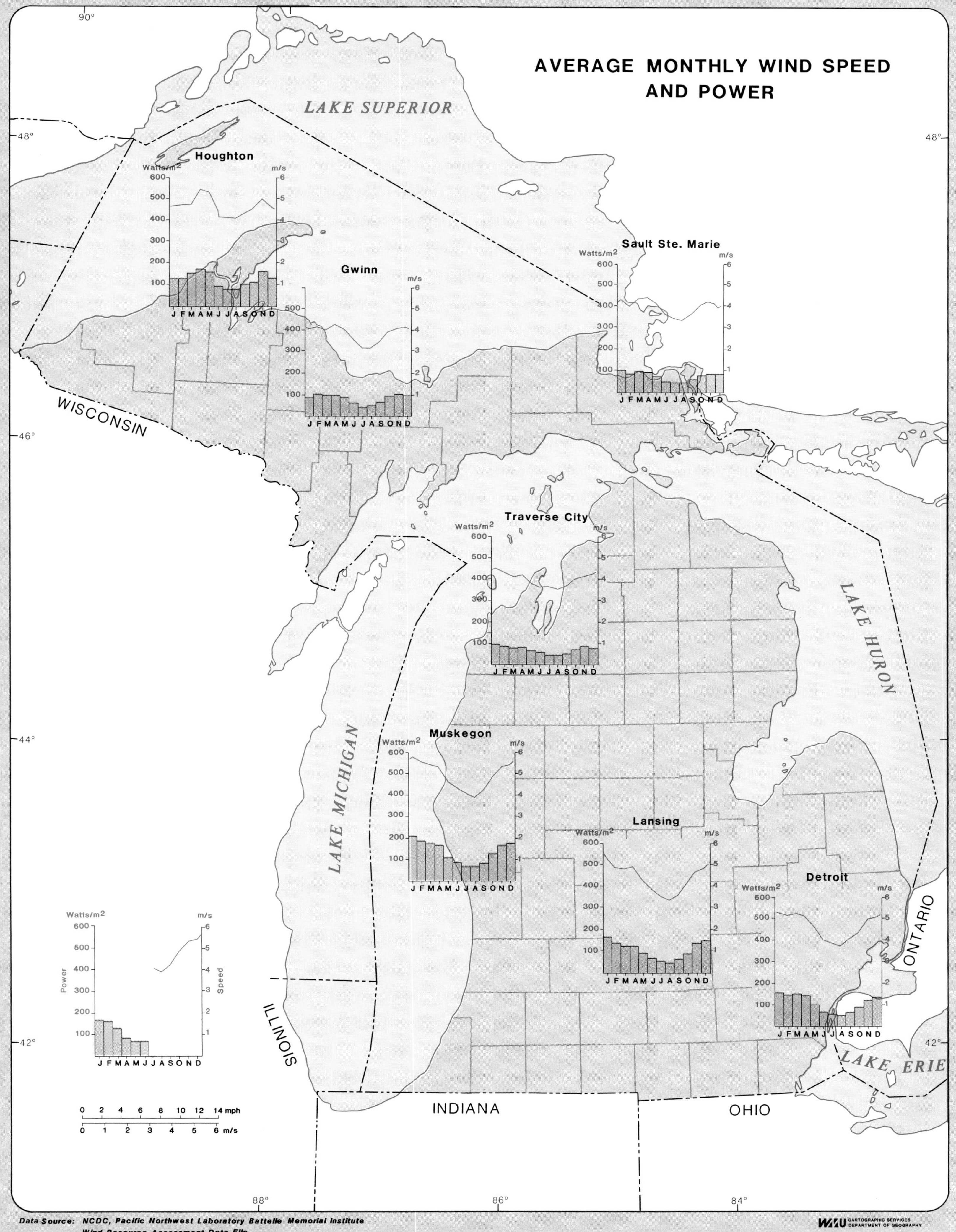


Data Source: NCDC, Pacific Northwest Laboratory Battelle Memorial Institute
Wind Resource Assessment Data File

WMU CARTOGRAPHIC SERVICES DEPARTMENT OF GEOGRAPHY

## Average Annual Diurnal Wind Speeds

Maximum wind speeds occur between noon and 3:00 p.m., or during the afternoon when surface heating is at a maximum and turbulence is greatest. The diurnal variation is largest in summer and least in winter when surface heating during the day is minimal.

## Average Annual Wind Speed Frequency

Winds at the majority of the stations have peak frequencies at speeds of about 4 mps (8.9 mph). The peak frequency at Sault Ste. Marie is 3 mps (6.7 mph). Observer bias may play a role in causing the apparent sharp changes of frequency from one speed to the next.

## Average Annual Wind Speed Duration

The graphs show the percent of time the indicated wind speed value is equaled or exceeded. Sharp changes in percent from one speed to the next may again indicate some bias in wind observations.

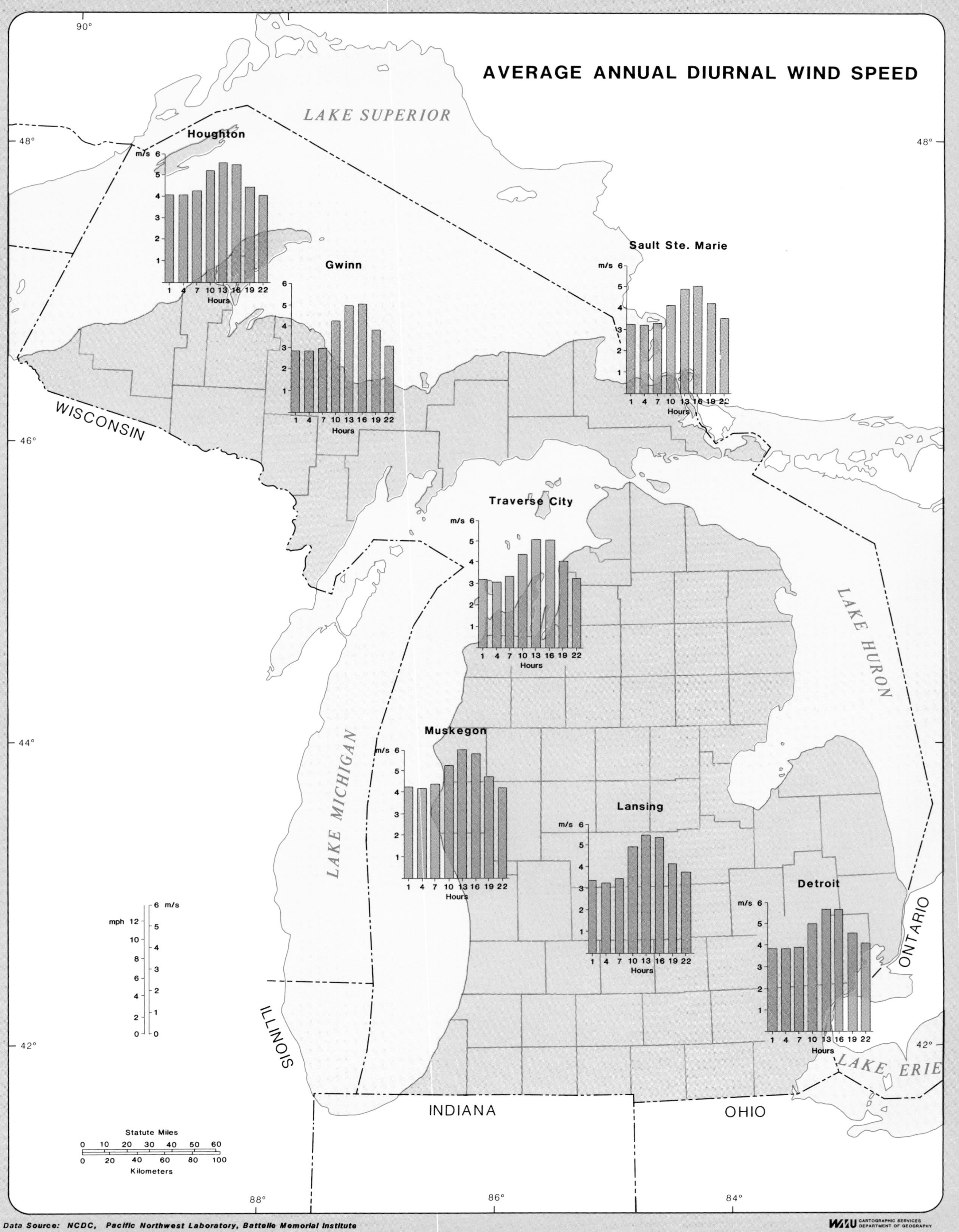


Data Source: NCDC, Pacific Northwest Laboratory, Battelle Memorial Institute Wind Resource Assessment Data File

WMU Cartographic Services Department of Geography

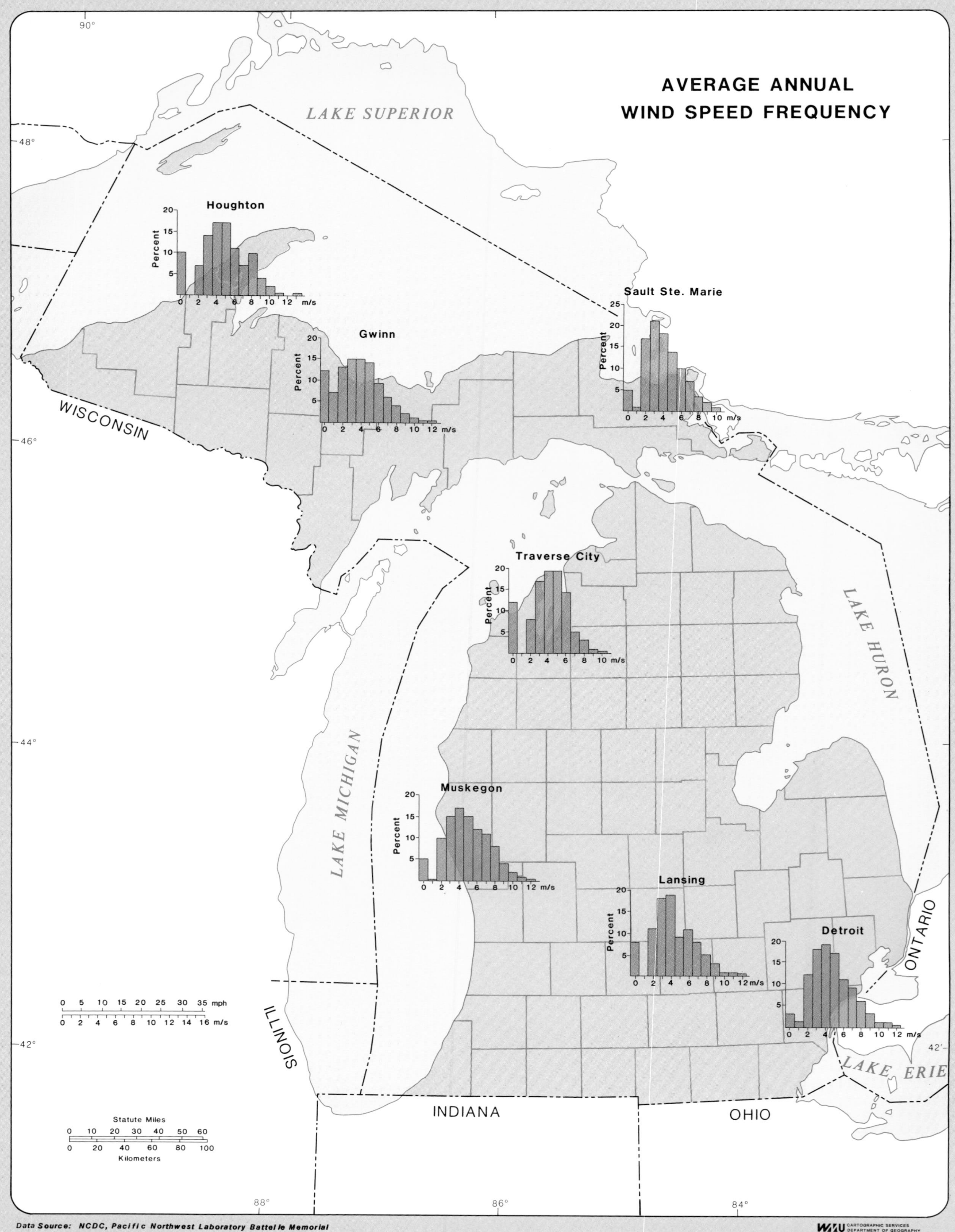


Data Source: NCDC, Pacific Northwest Laboratory Battelle Memorial Institute Wind Resource Assessment Data File

WMU CARTOGRAPHIC SERVICES DEPARTMENT OF GEOGRAPHY

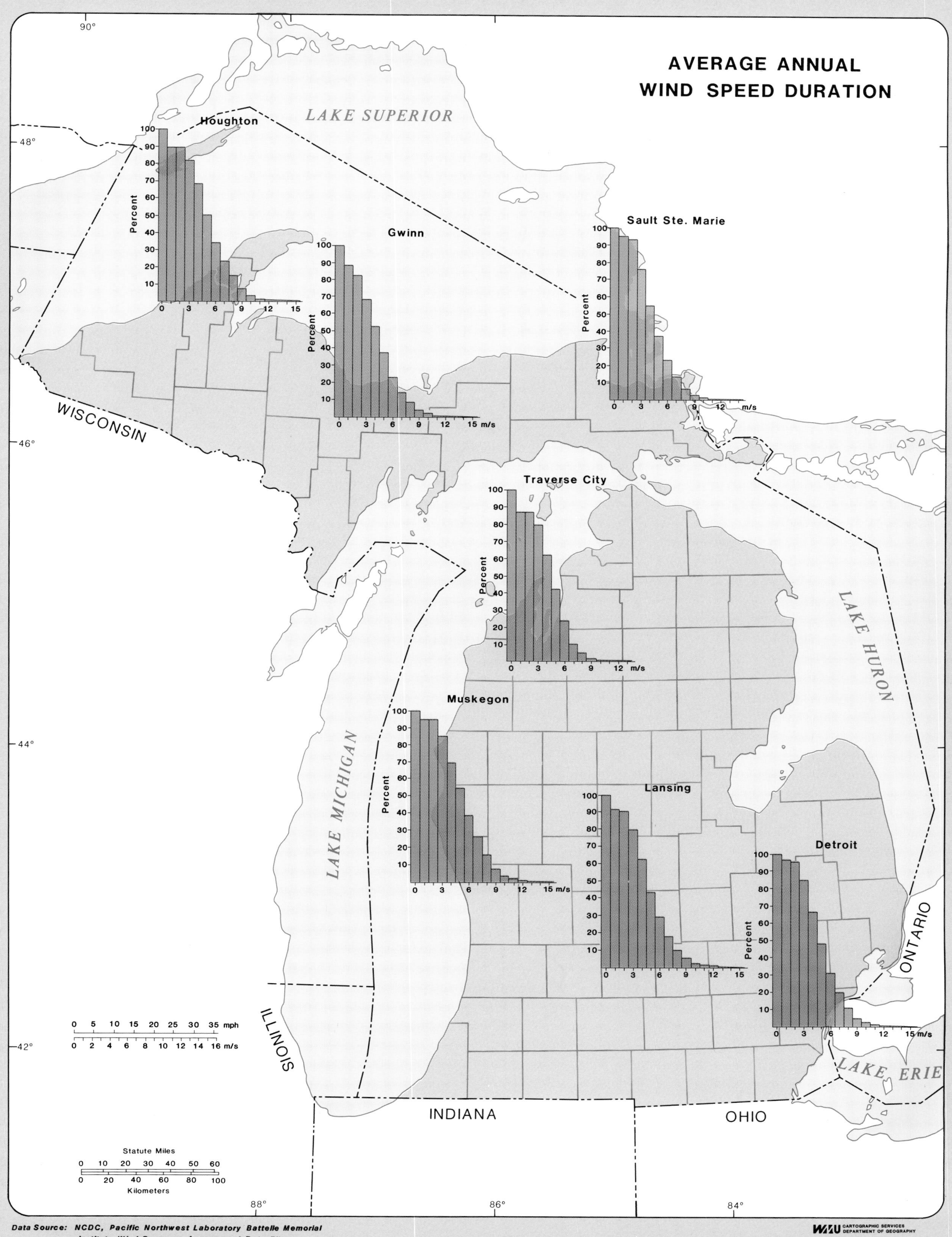


Data Source: NCDC, Pacific Northwest Laboratory Battelle Memorial Institute Wind Resource Assessment Data File

WMU CARTOGRAPHIC SERVICES DEPARTMENT OF GEOGRAPHY

# SUNSHINE AND CLOUDINESS

Unlike some parts of the United States, Michigan experiences a wide variation throughout the year in sunshine and cloudiness. Summers are relatively sunny, but sunshine is rare in the winter and cloudiness excessive. This marked seasonal contrast is caused by the southward shift of the westerlies and polar front in winter, bringing an increased frequency of low pressure areas. In addition, the heat and moisture from the Great Lakes transferred into the atmosphere during the cold season increases the cloud cover, making Michigan one of the cloudiest areas in the United States in the winter.

Observations of cloud cover are made visually at a number of National Weather Service stations in Michigan. The monthly number of clear, partly cloudy, and cloudy days and the percent cloud cover (not shown on the graphs) are derived from the visual estimates. Data indicating the percent sunshine, however, are derived from instrumental observations. The Maring-Marvin thermoelectric sunshine duration recorder was used in Michigan until the 1950s and 1960s when the instrument was supplanted by the Foster-Foskett photoelectric sunshine switch. The percent cloudiness or total cloud cover is not the reciprocal of the percent sunshine because the cloud cover is overestimated by ground-based visual observations due to line-of-sight problems and the sunshine sensors may fail to respond to high thin clouds. Thus the instrumental estimates of cloud cover have been about 15% lower than the visual observations.

Mean monthly values for percent sunshine and cloudiness, and number of clear, partly cloudy, and cloudy days are shown for Alpena, Detroit, Grand Rapids, Lansing, Marquette, and Sault Ste. Marie. The sources for the data were the publications *Climatological Data—Michigan* and the *Monthly Weather Review*. The time period for the data is 1923–1979.

The graphs indicate the strong seasonal contrast in sunshine and cloudiness in the state. Percent sunshine ranges from a minimum value of 23% at Sault Ste. Marie in November to maximum of 73% at Lansing in July. In general, sunshine is less and cloudiness greater at more northerly locations throughout the year, except in spring.

# AVERAGE MONTHLY SUNSHINE AND CLOUD COVER 1923-1979

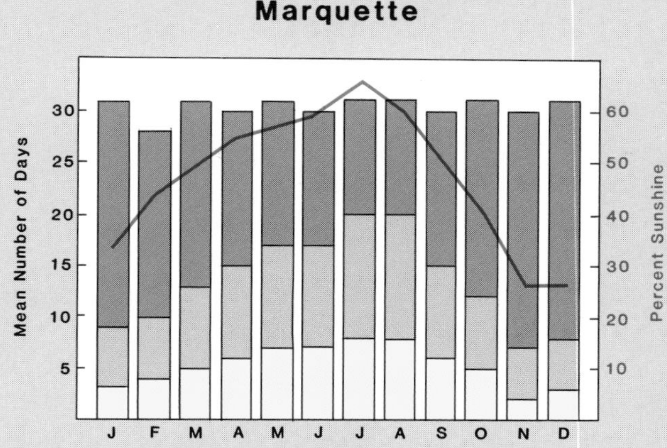


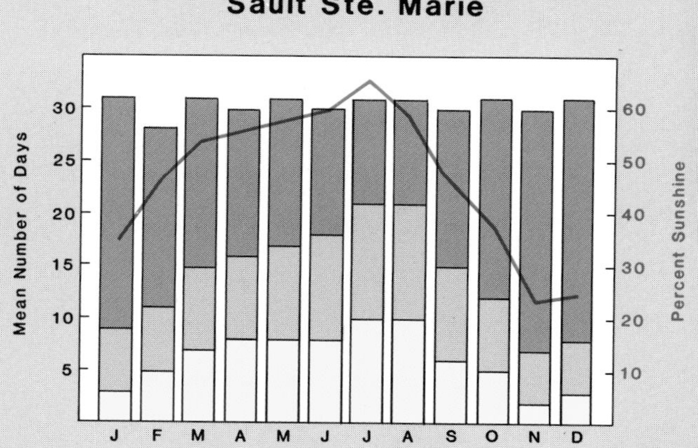


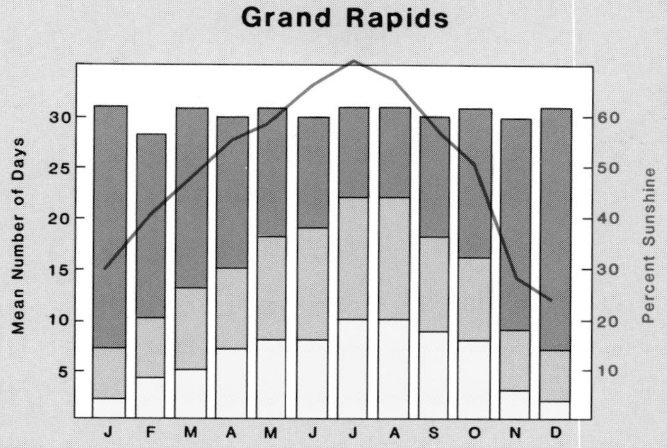


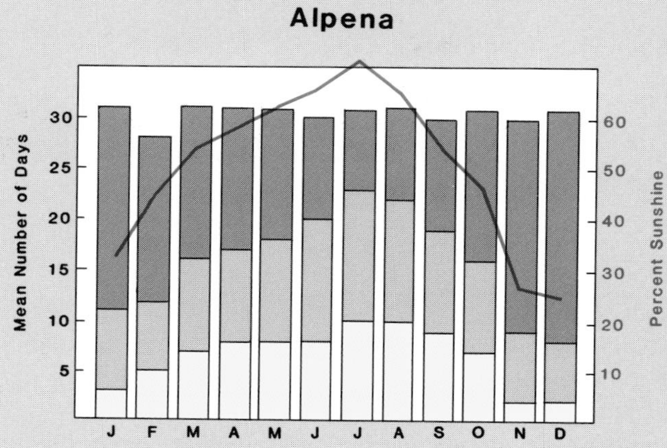


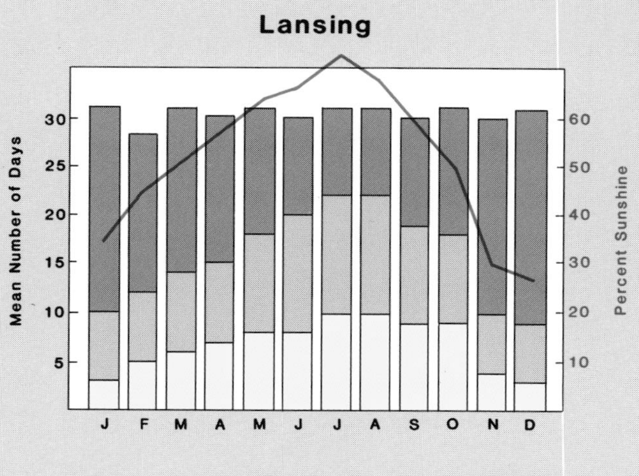


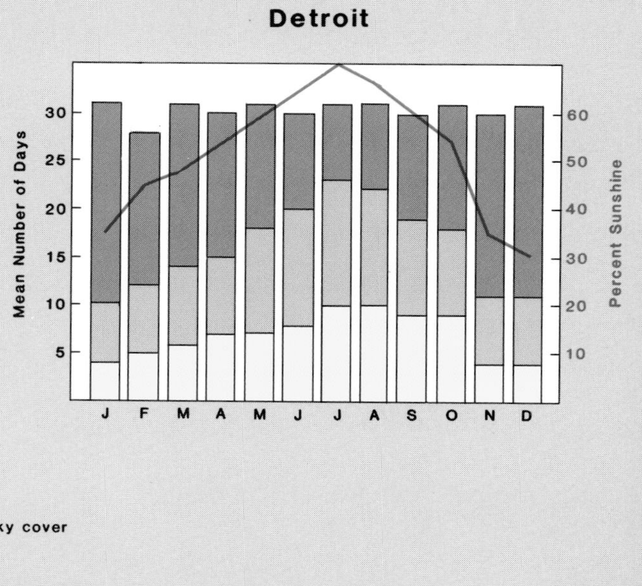


Cloudy Days 0.8–1.0 sky cover

Partly Cloudy Days 0.4–0.7 sky cover

Clear Days 0.0–0.3 sky cover

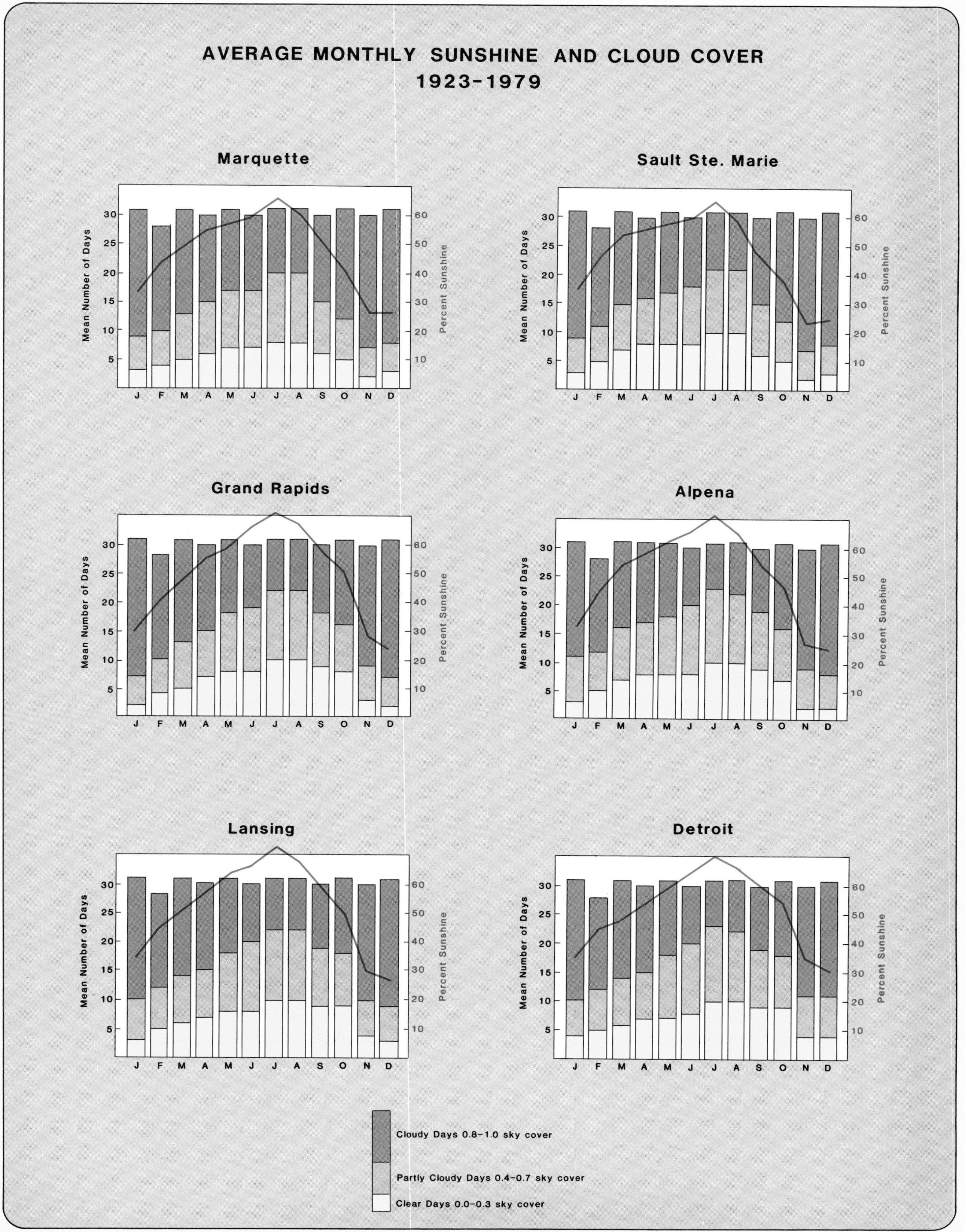

*Data Source: Climatological Data - Michigan*

WMU CARTOGRAPHIC SERVICES DEPARTMENT OF GEOGRAPHY

# SOLAR RADIATION

Michigan sunset.

Solar radiation is the source of energy for atmospheric processes and systems. It is also the source of nearly all energy utilized by humans, primarily as a key element in the formation of fossil fuels. More recently, as solar energy is inexhaustible, its direct usage has been viewed as a future alternative to fossil fuels.

Unfortunately, accurate and usable solar radiation data are scarce because of a prevalence of errors due to problems with instrumentation, data collection, and presentation. Uncorrected and unedited data have been available for only two Michigan stations, Sault Ste. Marie and East Lansing.

The National Climatic Data Center (NCDC), in cooperation with a number of other agencies, has provided regression-modeled data for 211 stations in the United States (SOLMET and ERSATZ data). Among these have been six Michigan stations, Alpena, Detroit, Flint, Grand Rapids, Sault Ste. Marie, and Traverse City. The solar radiation values for these stations have been estimated from more commonly reported data including amount of cloudiness, minutes of sunshine, and solar angle. Although the data will be continuously evaluated for accuracy, they are, at least for the time being, the most valid data available.

The data are presented as graphed percentiles of mean daily total (global) radiation received at the earth's surface for seven-day periods throughout the year. The data values are given in kilojoules/$m^2$. The values may be converted to BTU/$F^2$ by multiplying by $8.811 \times 10^{-2}$. The time period of the data is 1952–1976.

The seasonal variation of solar radiation in Michigan is a response to the seasonal variations in the solar altitude and length of day. At Sault Ste. Marie, the most northerly of the stations shown, the noon solar altitude varies from about 20 degrees at the winter solstice to about 67 degrees at the summer solstice. The length of day varies from about 8 hours and 32 minutes to about 15 hours and 49 minutes. At Detroit, the most southerly of the stations, the noon solar altitude varies from about 24 degrees to about 71 degrees, and the length of day from about 9 hours and 04 minutes to about 15 hours and 17 minutes. Solar radiation values increase as does the solar altitude and the length of day.

In addition, the amount of cloud cover affects the daily total of solar radiation. As Michigan exhibits large seasonal contrasts in cloud cover (see Sunshine and Cloudiness Section), this further increases the seasonal differences in solar radiation reception.

The largest values at the 50th percentile level occur during the 23rd and 26th seven-day periods (June 4–10 and June 25–July 1). Values range from 25,630 kJ/$m^2$ at Grand Rapids during period 23 to 23,552 kJ/$m^2$ at Flint during period 26.

The smallest values occur during the 51st and 52nd periods (December 17–23 and December 24–31). Values range from 3319 kJ/$m^2$ at Detroit during period 51 to 2079 kJ/$m^2$ at Sault Ste. Marie during period 52.

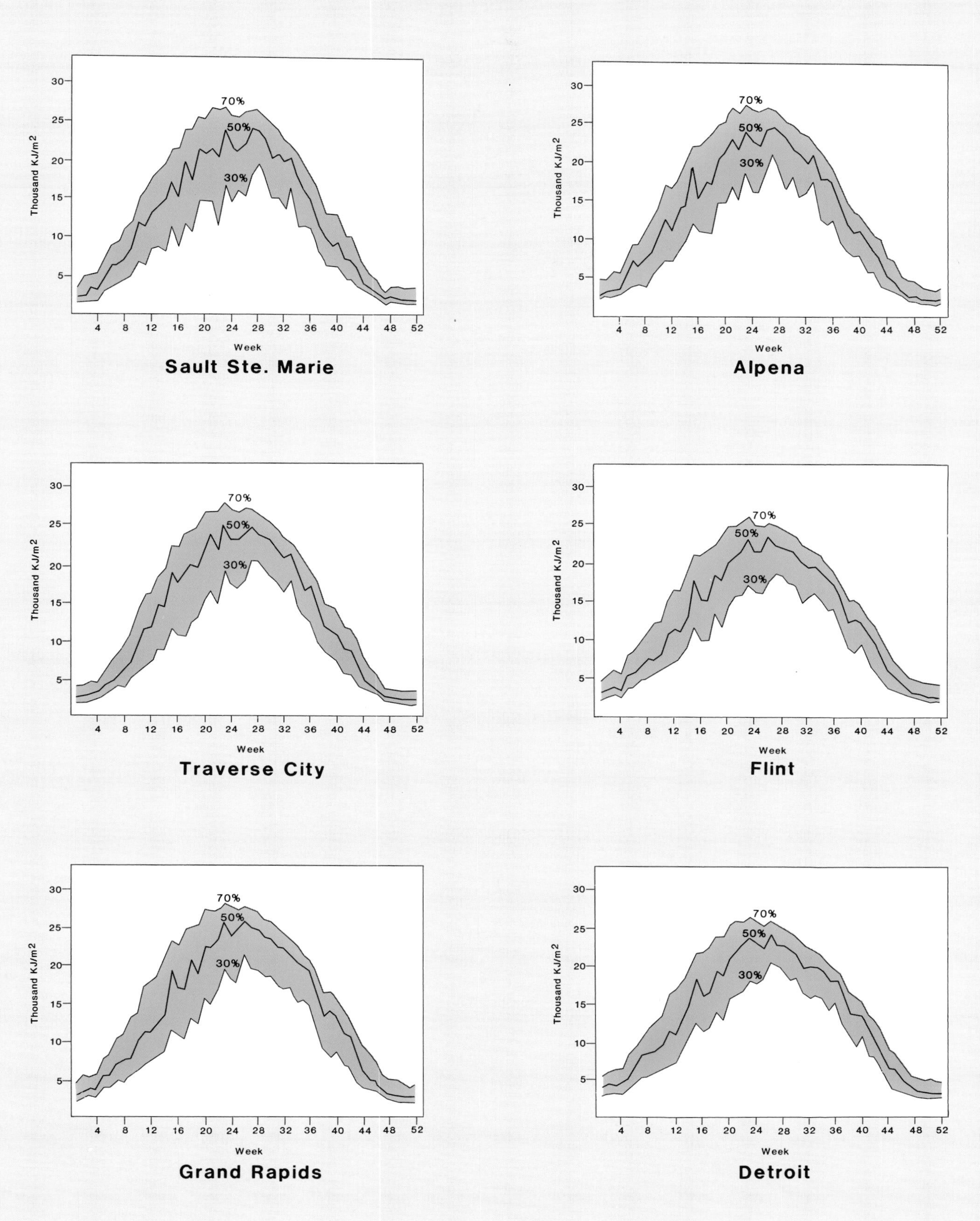

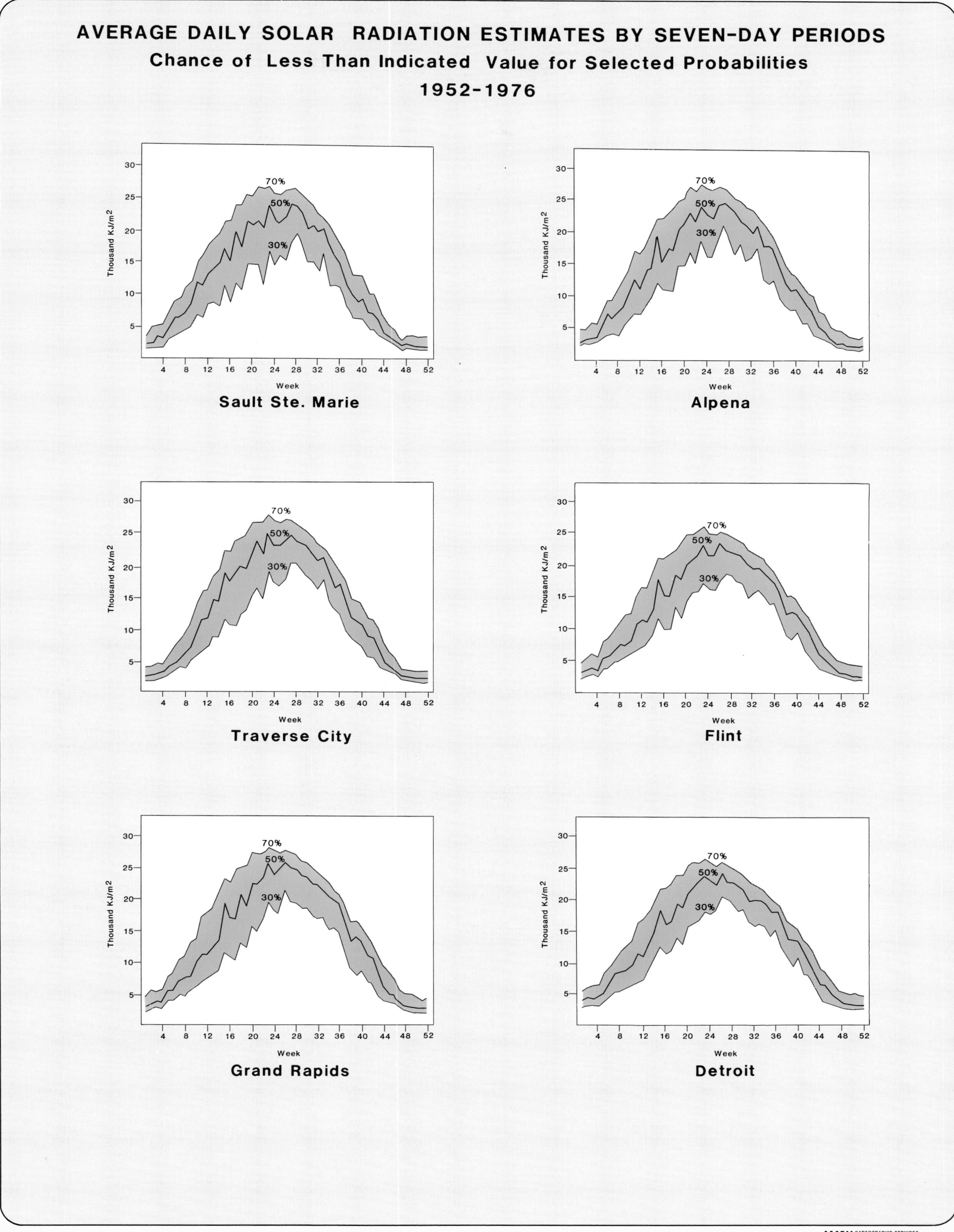

Source: V.L. Eichenlaub, Solar Radiation Estimates for Six Michigan Stations
(Data from SOLMET data tapes provided by NCDC)

WMU CARTOGRAPHIC SERVICES DEPARTMENT OF GEOGRAPHY

# PRESSURE

Atmospheric pressures in Michigan vary considerably from day to day in response to migratory high and low pressure systems (anticyclones and cyclones). Temperature variation also affects pressure readings. Cold air is more dense than warm air and pressures tend to be higher when temperatures are low and lower when temperatures are high.

The actual annual pattern of average monthly pressure in Michigan is complex, reflecting both seasonal temperature variation as well as seasonal changes in the frequencies of high and low pressure systems. Low pressure systems cross the state with great frequency during the cold season, while high pressure systems occur infrequently at that time of year. During the warm season, the opposite is true.

Average monthly station pressures are shown for six stations, Sault Ste. Marie, Alpena, Houghton Lake, Muskegon, Lansing, and Detroit Metropolitan Airport. The data source was the publication *Local Climatological Data—Annual Summary*. The monthly pressures indicated by the bar graphs are not adjusted for differences of altitude of the six stations, and hence the values are not directly comparable and are lower than those reported by the news media, which are adjusted to mean sea level. The period of record is thirteen years, except for Houghton Lake, where only seven years of data were available.

The annual pattern of pressure change at the six stations shows, in general, that highest average pressures occur in the late summer and fall, when temperatures are cooler and anticyclones still traverse the state with some frequency. A secondary peak occurs during the winter cold months. Lowest average pressures occur in the spring when temperatures are warmer and traveling cyclones still frequently cross the Midwest. The annual range of monthly average pressure is not large, ranging between 2 and 4 millibars (about .06 and .12 inches of mercury).

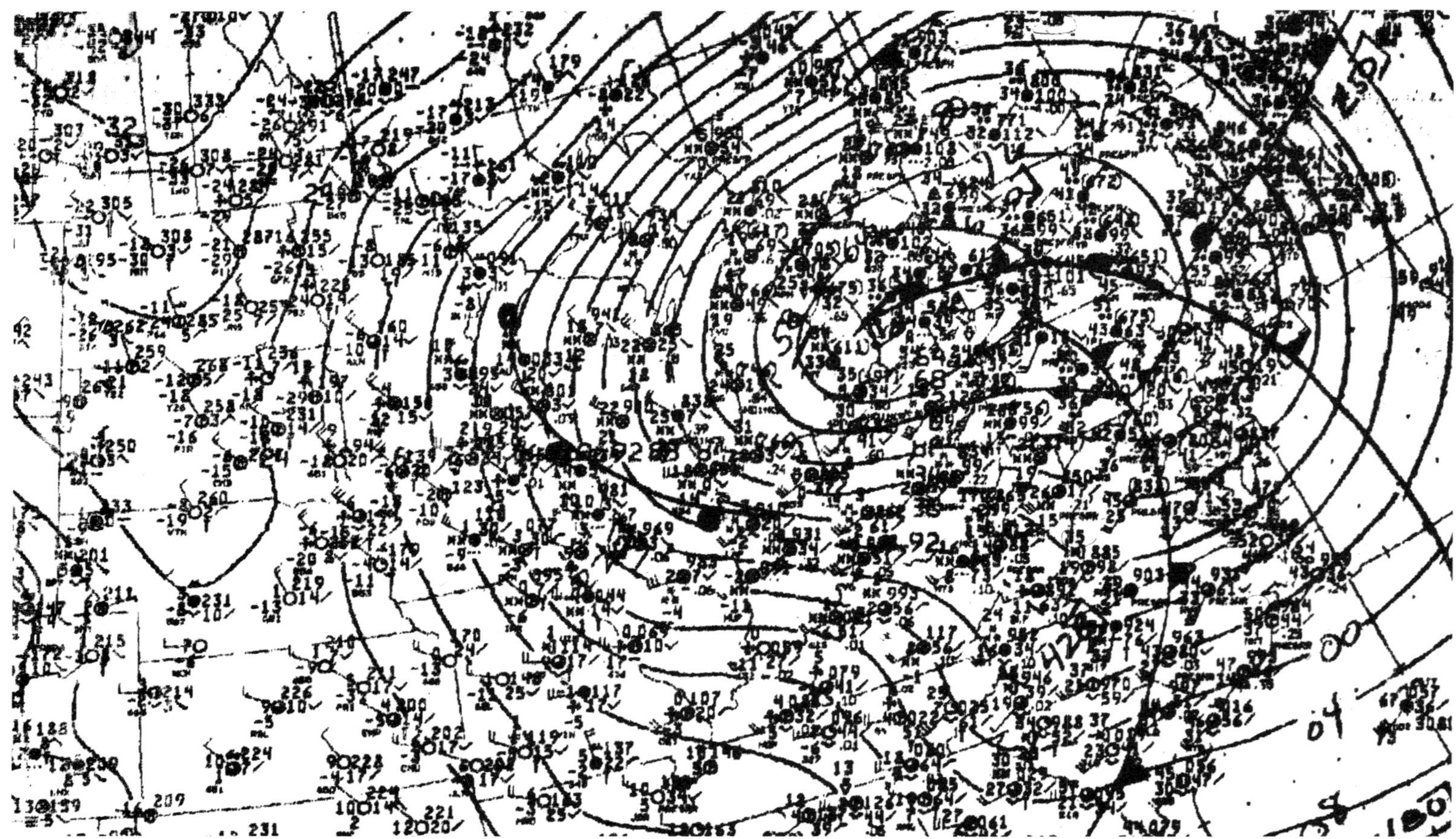

Surface map 7:00 a.m. EST, January 26, 1978. Storm described as "worst blizzard of century" for parts of Michigan. Surface pressures of 958 mb were the lowest ever recorded in the Great Lakes region.

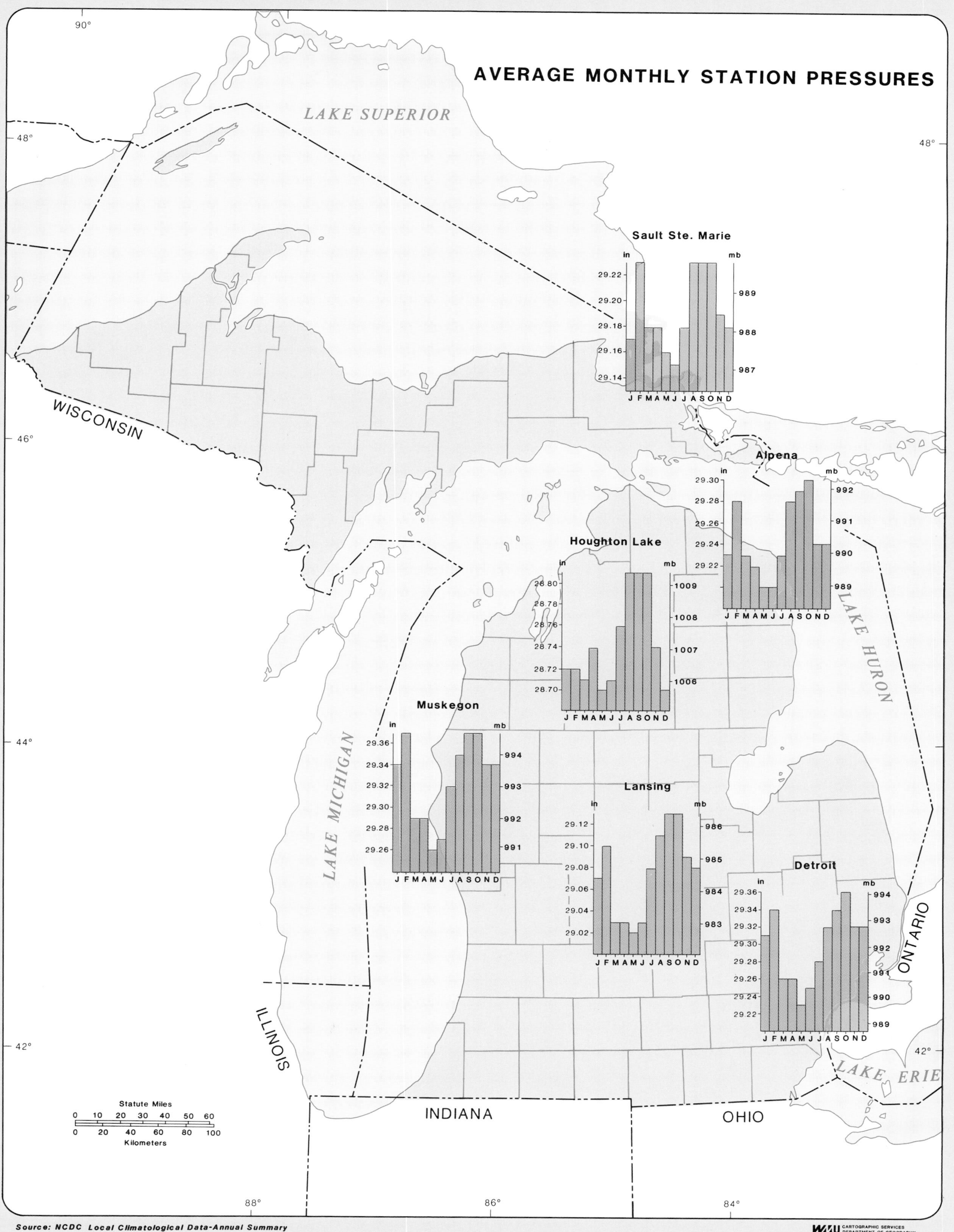


Source: NCDC Local Climatological Data-Annual Summary

WMU CARTOGRAPHIC SERVICES DEPARTMENT OF GEOGRAPHY

# FRUIT BELT CLIMATE

Commercial fruit production is an important agricultural activity in the counties of the western Lower Peninsula from the vicinity of Grand Traverse Bay southward to the Indiana state line. In this "fruit belt," peaches, grapes, apricots, pears, cherries (both sweet and tart), plums, and apples are raised on a scale exceeding that of any other region of the state. Most of these fruits are relatively sensitive to extreme low temperatures at several stages in their annual growth cycle. Their commercial production is clustered on the east side of Lake Michigan where the prevailing winds across the lake help ameliorate these extremes.

Because most of these plants set fruit from flower buds formed the previous summer, they are particularly vulnerable to injury by extreme cold during the winter, and geographic patterns of minimum temperatures explain in large part the location of the fruit belt in the western Lower Peninsula. The accompanying maps showing the number of occurrences of temperature at or below the specified values reveal the greatly reduced total frequency of low temperature extremes nearer the lake during the 1951–1980 period. This effect results from either lake-induced cloudiness or the direct modification of cold air moving across the lake into the state from the west. As expected, 0°F values are observed most frequently over the entire area, and values of −15°F or lower are much less frequent. On all three maps the zone of lower frequency near the lake narrows toward the north, implying that lake-induced climatic modification becomes much more restricted.

Maps of the number of years with minimum temperatures at or below the specified values reveal that the overall frequency of these temperatures increases northward and inland. As with the previous maps, the northward increase is least developed near the lake, again suggesting that the lake moderates the local winter climate and reduces the frequency of extremes of cold that might be detrimental to sensitive fruit.

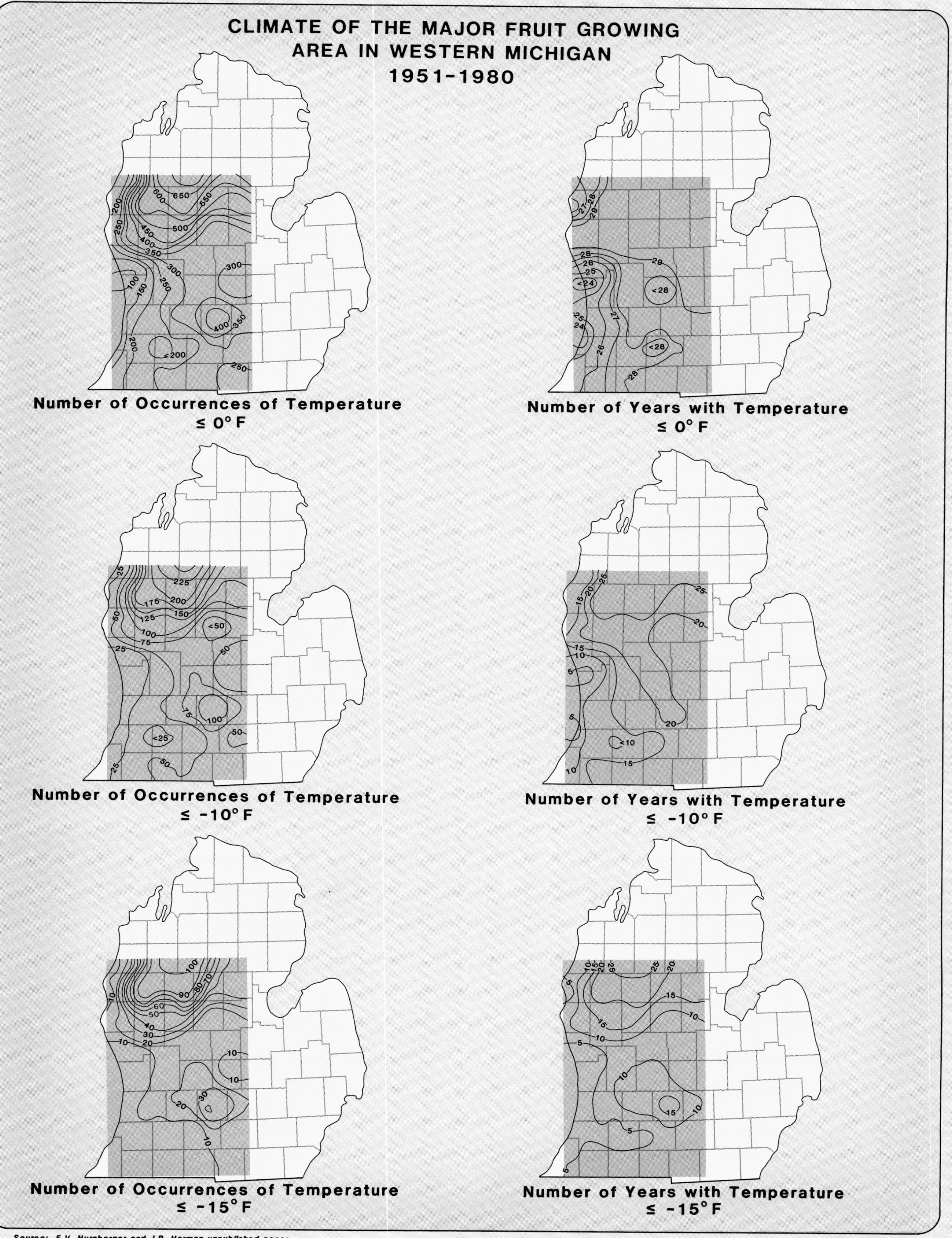


*Source: F.V. Nurnberger and J.R. Harman unpublished paper*

WMU CARTOGRAPHIC SERVICES DEPARTMENT OF GEOGRAPHY

# COMFORT CLIMATES

The feeling of comfort is a partially subjective response to climate. Although marked individual differences in the perception of comfort occur, comfort seems to be central to much of the human reaction to climate. It dictates our customs in housing, clothing, outdoor activities, and in many other aspects of our daily lives.

The depiction of comfort climates necessitates the use of combinations of individual weather elements. For example, it is well known that comfort during Michigan summers is a response not only to temperature, but also to humidity, sunshine, and wind. In winter, comfort is affected not only by temperature, but also by wind and sunshine. In addition, comfort is very responsive to diurnal and interdiurnal changes of weather and hence is difficult to express in terms of mean monthly values.

The graphs on the following pages show "comfort" conditions for January and July at Grand Rapids Kent County Airport for the years 1965–1968. No attempt has been made to depict these as "average" years, but, rather, the graphs give some idea of the range of comfort conditions which may occur and the interdiurnal variability which may be expected. The data for Grand Rapids can be assumed to give rough approximations of comfort conditions at other southern Michigan stations.

The graphs are derived from a study *Comfort Climates at Grand Rapids, Michigan* by Robert L. Janiskee using the Terjung physioclimatic classification scheme as presented in Werner H. Terjung, "Physiological Climates of the Conterminous United States." With this scheme specific Comfort Groups derived from combinations of temperature, humidity, wind, and sunshine are identified. These are thought to describe the comfort sensations of an average individual.

## January Comfort Groups

A total of 992 tri-hourly observations for the four Januarys, 1965–1968, entered into the determination of 5 Comfort Groups. The frequency of occurrence and interdiurnal variability of the Comfort Groups for the four Januarys of the study period are shown on the graphs.

The five Comfort Groups occurring during daytime in January and their percentage of total occurrences were as follows:

| | | |
|---|---|---|
| C | Cool | 0.5% |
| K | Keen | 18.5% |
| CD | Cold | 66.9% |
| VC | Very Cold | 13.7% |
| EC | Extremely Cold | 0.3% |

The groups are ranked from least uncomfortable to most uncomfortable with only the cool group regarded as truly comfortable.

Comfort Groups occurring at night and their percentages of total occurrences in January include:

| | | |
|---|---|---|
| C | Cool | 0.2% |
| K | Keen | 13.4% |
| CD | Cold | 63.7% |
| VC | Very Cold | 21.0% |
| EC | Extremely Cold | 1.8% |

During the nighttime, a shift toward the Very Cold Comfort Group occurs as this group has a higher frequency than the Keen Comfort group.

The interdiurnal variability measures the day-to-day variability of the *warmest* daily Comfort Group. Consecutive occurrences of days with the same warmest Comfort Group value are called Comfort Spells. Short Cycle Intervals represent brief intervals of Comfort Spells or transitional periods. Keen Spells are perceived as relatively mild daytime comfort conditions, but still on the cool side for comfort. Cold Spells represent normal daytime comfort conditions for January. Very Cold Spells are indicative of unusually cold daytime comfort conditions. Each month is seen to consist of 5 or 6 Comfort Spells and 2–4 Short Cycle Intervals. This suggests that Michigan Januarys contain spells of varying discomfort, but the days are nearly always too cold for comfort.

# JANUARY COMFORT PROFILES 1965-1968

## Grand Rapids

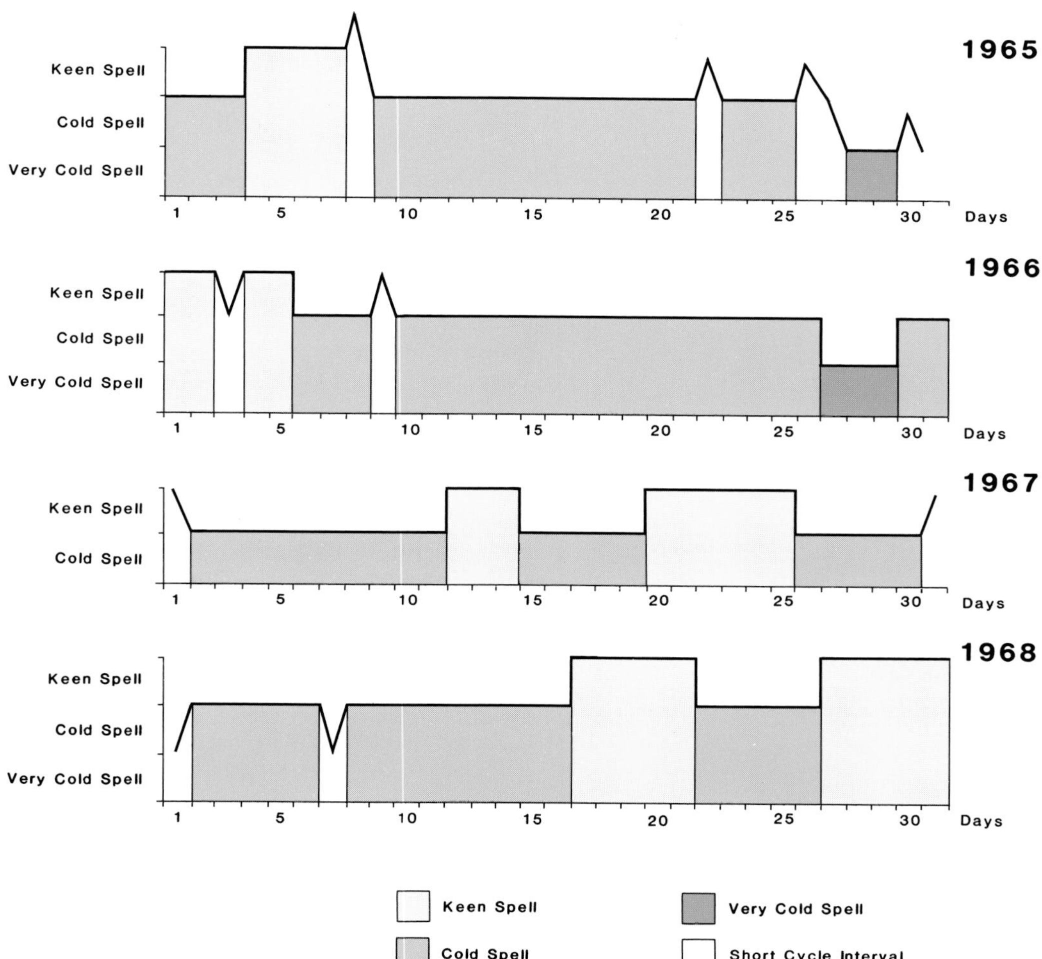


# FREQUENCY OF JANUARY COMFORT GROUPS

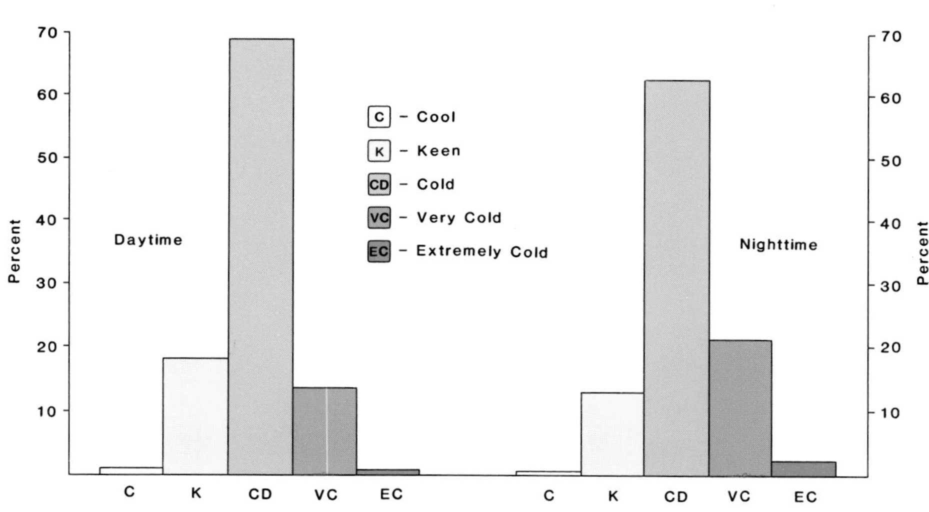


Source: R.L. Janiskee Comfort Climates of Grand Rapids, Michigan

WMU CARTOGRAPHIC SERVICES DEPARTMENT OF GEOGRAPHY

# July Comfort Conditions

The frequency of occurrence and interdiurnal variability of Comfort Groups for the four Julys of the study period are shown on the graphs. A total of six Comfort Groups were derived from the 992 tri-hourly observations. Mild and Cool were considered the most comfortable.

The six Comfort Groups and their frequency of occurrence during daytime in July included:

| | | |
|---|---|---|
| S | Sultry | 5.2% |
| H | Hot | 5.2% |
| W | Warm | 31.0% |
| M | Mild | 42.1% |
| C | Cool | 8.4% |
| K | Keen | 8.2% |

Daytime July observations ranged from much too hot to too cool, but also included a large number of occurrences of conditions that were comfortable for the average person.

Five Comfort Conditions occurred at night:

| | | |
|---|---|---|
| S | Sultry | 0.5% |
| W | Warm | 8.9% |
| M | Mild | 38.2% |
| C | Cool | 21.5% |
| K | Keen | 30.9% |

Over 90% of the observations occurred in the Mild to Keen portions of the Comfort Groupings.

As with January, the classification of an individual day is based on the warmest *Comfort Group*. Recognized were: Hot Spells (2 or more consecutive days with Comfort Index H or S), Warm Spells (2 or more days with Comfort Index W), and Comfortable Spells (2 or more consecutive days with Comfort Index M or C). Single days were classed as Short Cycle Intervals. Hot Spells are undesirable and may cause substantial heat discomfort. They occurred occasionally during all the Julys of the study period. Warm Spells may cause some heat discomfort and occurred more frequently. Mild spells were most desirable for human comfort and also occurred frequently during all four of the Julys studied. This indicates that many July days in Michigan, unlike January days, are comfortable for the average person.

# JULY COMFORT PROFILES 1965-1968

## Grand Rapids

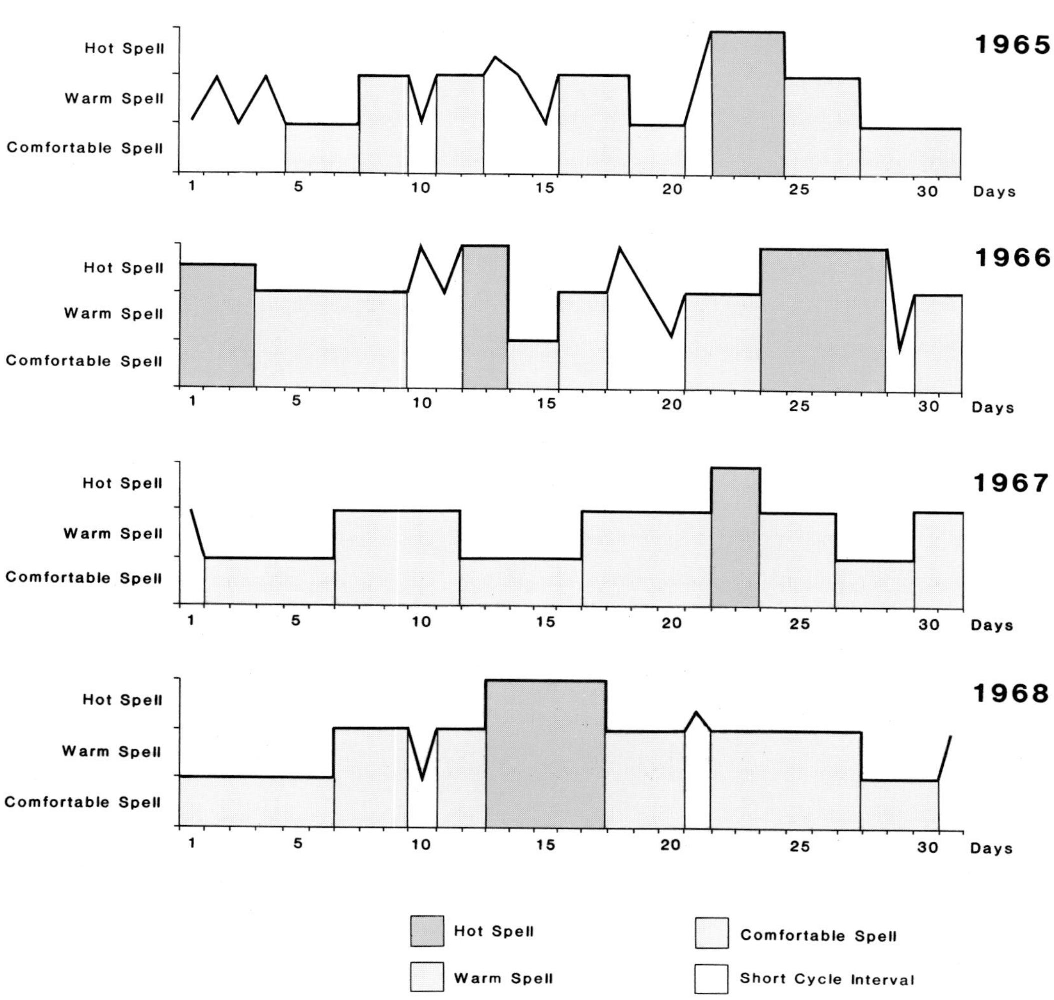


# FREQUENCY OF JULY COMFORT GROUPS

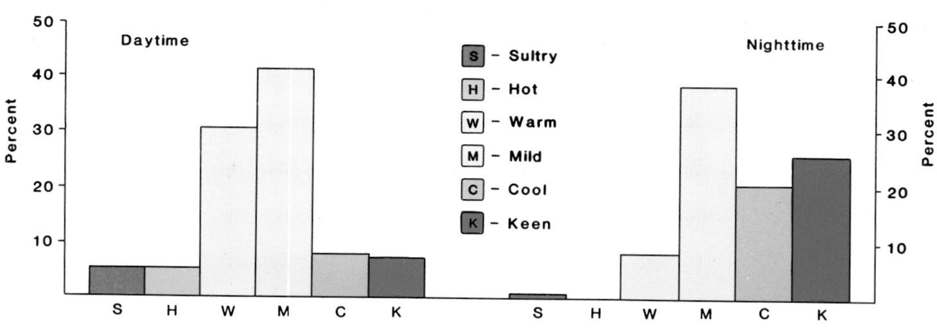


Source: R.L. Janiskee Comfort Climates of Grand Rapids, Michigan

WMU CARTOGRAPHIC SERVICES DEPARTMENT OF GEOGRAPHY

# CLIMATE CHANGE IN MICHIGAN

Jet contrail over southern Michigan. Some researchers feel the increase of jet traffic has been partly responsible for the observed decrease of sunshine over Michigan and much of the midwest.

Has Michigan's climate changed? Is it likely to change in the future? Questions such as these have become commonplace since the chaotic years of the 1970s when weather problems around the world caused massive crop failures, and bitterly cold winters in the eastern U.S. aroused speculation about the return of the ice ages.

Then, in the mid- and late 1980s, global temperatures rose to their highest values of the century. During the summer of 1988, heat and drought were proclaimed by the media to be harbingers of the "dreaded" greenhouse effect. Low water levels for the Great Lakes were projected for future years. Forgotten were the heat and droughts of the 1930s, for which a greenhouse effect could not be culpable. It was also forgotten that just two years prior, at the culmination of a series of inordinately wet years, high water levels for the Great Lakes were predicted for many decades to come.

Given these conflicting scenarios, it is necessary to look to the past and examine carefully the data for Michigan for the current century. In this way the scale of past climate fluctuations in the state can be appraised, providing a perspective within which to view abnormalities of current individual years or seasons. It is important also to recognize that climatic fluctuations that have occurred in Michigan need not be in phase or synchronus with global or hemispheric changes. Regional climatic changes may, in fact, occur in opposition to those at the hemisphere or global level.

## *Climatic Fluctuations in Michigan during the Twentieth Century*

Two well-marked climatic fluctuations have occurred in Michigan during the current century. One has been well documented over much of the United States and southern Canada. This is the progressive decrease in the range of daily temperature, described as "modulation" by researchers. The other distinctive change in Michigan has been more unique to the state and to the Great Lakes Region. This is the marked increase in lake effect snow which has occurred during recent decades of the twentieth century. These are discussed in V. L. Eichenlaub's article, "Some Recent Climatic Fluctuations in Michigan."

# Modulation of Daily Temperature Range in Michigan

In most parts of the United States the lessening of the annual average diurnal temperature range has resulted in the smallest values being reached in the 1980s. This trend, in Michigan, has been caused largely by a decrease in the average daily range of summer and fall temperatures. Winter and spring have not shown well-defined trends, although the range of spring daily temperatures dropped to smallest values of the century in the 1980s.

Figure 1 shows the decrease of July maximum temperatures and increase in minimum temperatures at Kalamazoo since 1930. In short, the average July day has become cooler (the blistering summer of 1988 being a notable exception) and the nights have become warmer.

Modulation has been attributed to a number of causes, but in Michigan it has been accompanied by (and probably caused by) a tendency toward decreasing sunshine and increasing cloudiness. Figure 2 shows the progressive dropoff at Grand Rapids in July in the percent of possible sunshine since the sunnier 1930s. This trend has been well marked, particularly in the summer, at most Michigan stations. More cloudiness prevents solar radiation from reaching the ground during the day, and prevents the loss of heat at night.

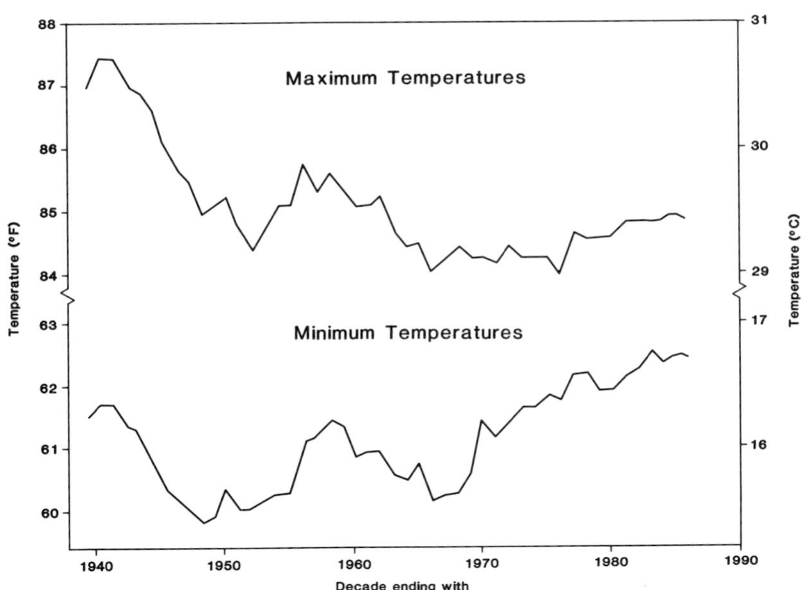


**FIGURE 1**
Ten-year moving averages of July maximum and minimum temperatures for Kalamazoo, 1930–1987. The moving average smooths the monthly values to give a better visual image of the trend.

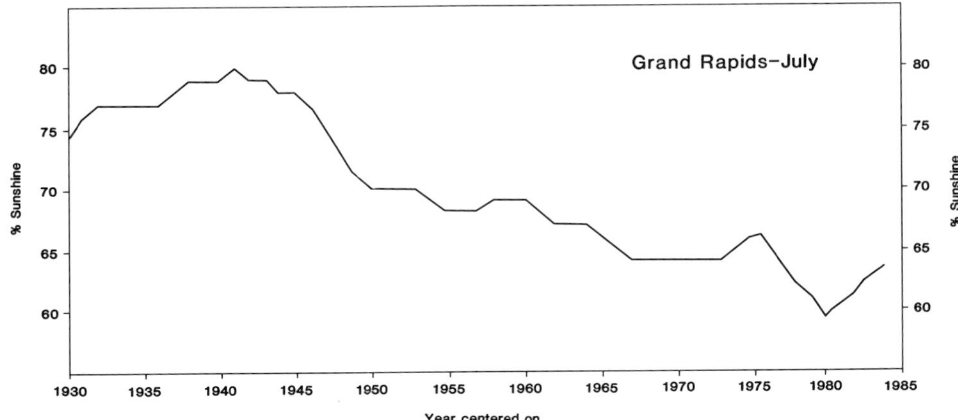


**FIGURE 2**
Percent of possible sunshine at Grand Rapids, 1926–1988, for July. The yearly values have been smoothed by a nine-point binomial filter to give a better visual image of the trend.

# Increases in Lake Effect Snowfall

A sharp increase in snowfall has occurred at many Michigan stations on the lee shores of Lake Michigan and Lake Superior. Figure 3 shows seasonal snowfall at Calumet-Houghton, on the Keweenaw Peninsula, 1887–88 through 1986–87. Snowfall averages have increased by nearly 100 inches since the early decades of the century. Surrounded by Lake Superior on three quadrants, this area is one of the snowiest in Michigan, receiving much of its total as lake effect snow.

Prior to April 1, 1948, the snowfall data for the Calumet-Houghton record were taken from a site at the Calumet and Hecla mining company offices, 0.5 miles south-southeast of the city of Calumet. Subsequently, the Houghton County Airport, 6 miles south-southeast of Calumet was designated as the official station. Thus, there is a lack of site homogeneity and observer uniformity and some caution must be used in interpreting the data. Nevertheless, as an upward trend began before the site change and continued afterwards, the increase appears to be real.

The record at Muskegon confirms that of Houghton-Calumet. At this station, where lake effect snow contributes a large part of the seasonal total, a large increase has occurred (figure 4). Seasonal snowfall averages during the 1960s and 1970s almost double those of the earlier decades of the century. At Lansing, where much less lake effect snow is normally received, the increases have been only moderate. It should be noted that, during recent years, seasonal snowfall totals appear to have decreased at Muskegon and stabilized at Houghton-Calumet.

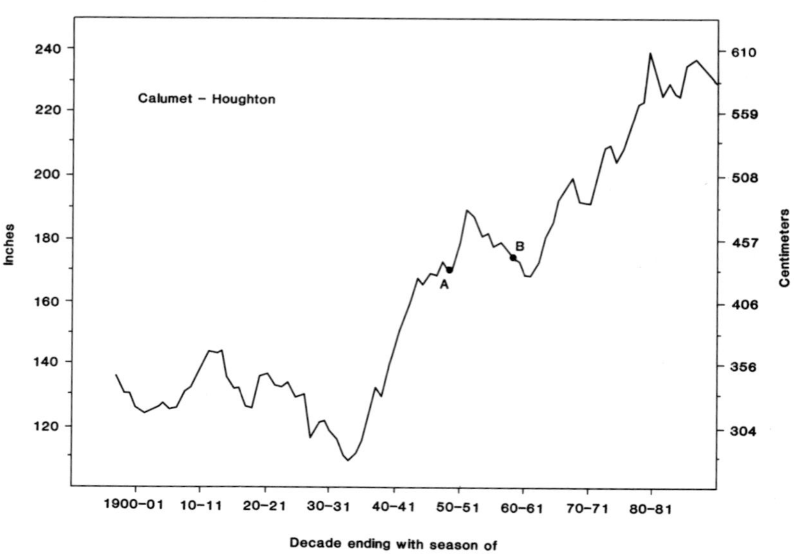


**FIGURE 3**
Ten-year moving averages of seasonal snowfall at Calumet-Houghton. A-B represents period in the data where a site change occurred.

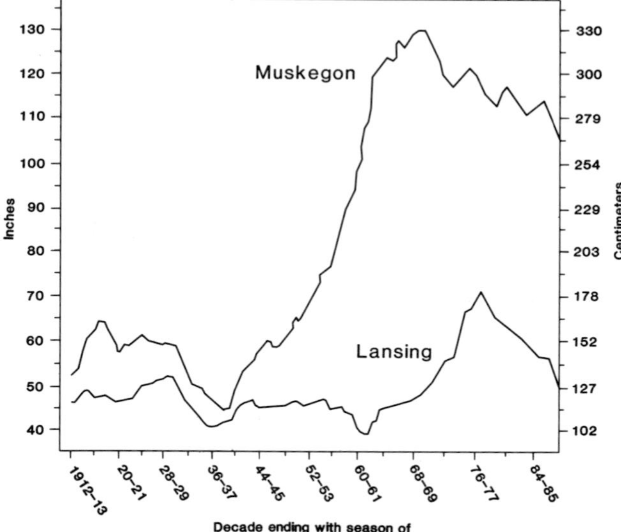


**FIGURE 4**
Ten-year moving averages of seasonal snowfall at Muskegon and Lansing, 1903–04 to 1986–87.

# Other Changes

## Temperature

Average annual daily temperatures, 1901–1984, have been plotted for a composite of stations in the Great Lakes region by Karl, Baldwin, and Burgin for the National Climatic Data Center. Included in the composite are ten stations from Michigan's Lower Peninsula and two from the Upper Peninsula. The graph (figure 5) shows a period from 1901 to about 1930 when annual temperatures were colder than the average. From the early 1930s to the late 1950s temperatures were warmer than the period average. From the late 1950s to the present, temperatures were near or a little below the average, reaching lowest values in the 1970s. A sharp upturn occurred in the 1980s.

The largest impact on annual temperatures in Michigan has been made by the individual seasons of winter and summer, particularly the peak seasonal months of January and July. Spring and fall have shown little change. The average daily mean temperatures, January and July, 1914–1987 are shown in figures 6 and 7 for Kalamazoo. January temperatures were colder than the

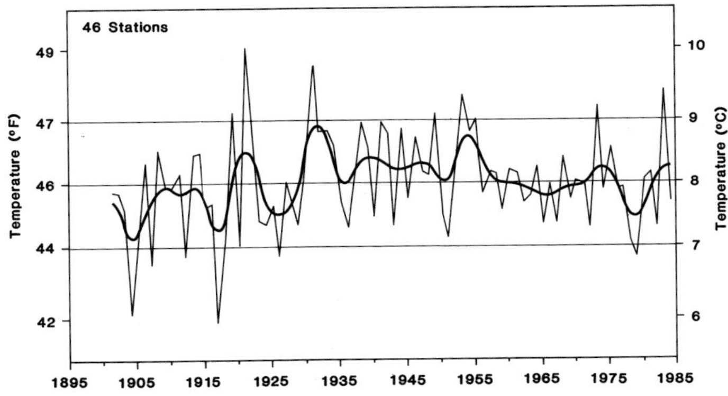


**FIGURE 5**

Average annual daily mean temperature for the Great Lakes Region (from Karl et al., *Time Series of Regional Season Averages of Maximum, Minimum, and Average Temperature and Diurnal Temperature Range Across the United States: 1901–1984*). Curve smoothed by nine-point binomial filter. Horizontal lines show 10th, 50th, and 90th percentile values.

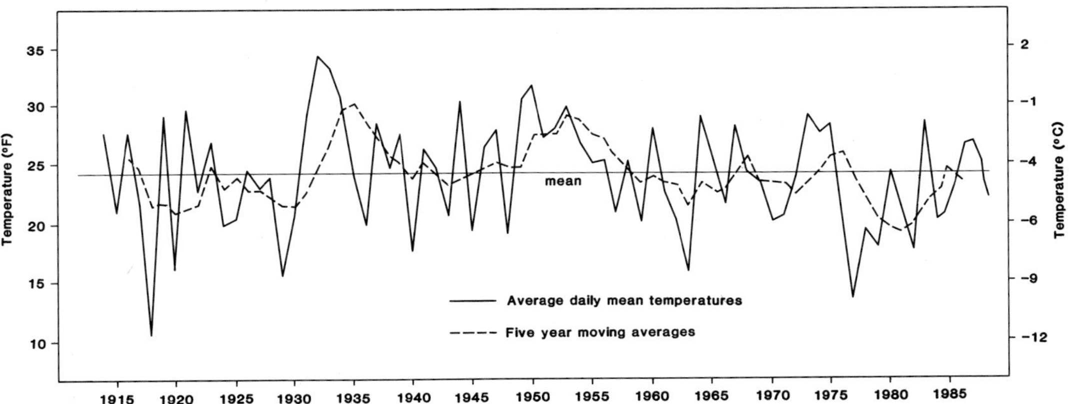


**FIGURE 6**

Average daily mean temperatures, January, 1914–1987, Kalamazoo. Dashed line shows five-year moving averages centered on year. Data prior to 1984 from an unpublished study by H. David Cole.

period mean prior to 1930, but warmer than the mean from about 1930 to the late 1950s. After that, Januarys became colder than the mean, with lowest values occurring in the late 1970s. An upturn during the 1980s has brought the values of the five-year moving averages only to the mean for the entire period. The cold Januarys during the 1960s and 1970s were probably largely responsible for the peak amounts of lake effect snowfall during those decades. With more cold air, the lake-atmosphere interactions which produce lake effect snowfall were intensified.

July average daily mean temperatures at Kalamazoo (figure 7) are more persistent around the period mean, although the warmer period of 1930 to the late 1950s is evident. It should be remembered that the decrease of average daily maximum temperatures in July which accompanied modulation is not depicted by the graph of average daily mean temperatures.

### *Precipitation*

The drought of 1988 was much publicized and had a major impact on Michigan's economy. However, records for the twentieth century show, during recent decades, increasing frequencies of "wet years" and decreasing frequencies of "dry years". In a study, "Climate Fluctuations and Record-High Levels of Lake Michigan" by S. A. Changnon, Jr., annual precipitation amounts for Michigan for the period 1896–1985 were ranked into three categories: annual precipitation values in the upper third were ranked as "wet"; values in the middle third were ranked as "normal", and values in the lower third were ranked as "dry". The frequency of "wet" and "dry" years were then plotted (figure 8). The results show that prior to 1941–1955, dry years were more frequent than wet years, but subsequently wet years have exceeded dry years.

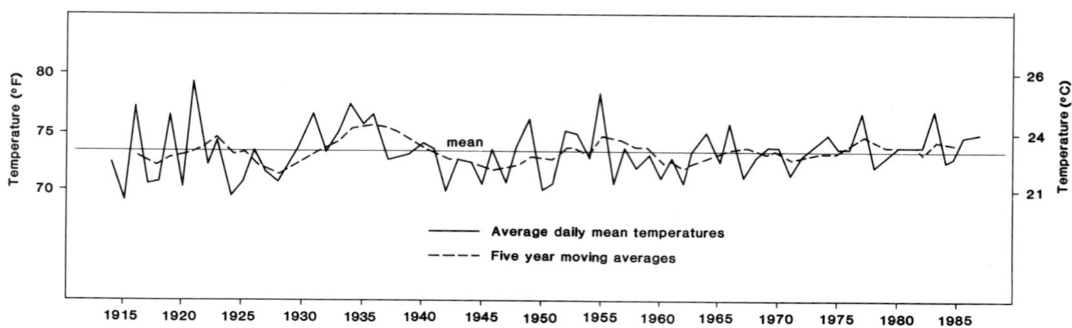


**FIGURE 7**
Average daily mean temperatures, July, 1914–1987, Kalamazoo. Dashed line shows five-year moving averages centered on year. Data prior to 1984 from an unpublished study by H. David Cole.

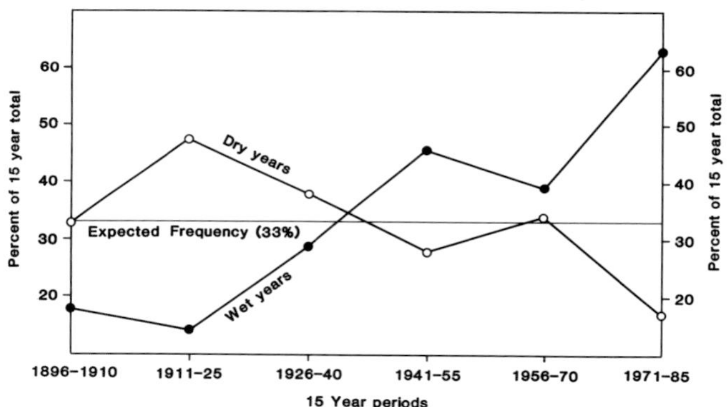


**FIGURE 8**
Frequencies of wet and dry years per fifteen-year period for Michigan (from S. A. Changnon, Jr., "Climate Fluctuations and Record-High Levels of Lake Michigan").

# Summary

Climatic fluctuations that have occurred in Michigan during the current century may be summarized as follows:

(1) The diurnal range of temperature has decreased, particularly in the summer. A likely cause is the increase in cloudiness at practically all stations for most months of the year.

(2) An increase in snowfall at lee shore locations has occurred. This increase appears to have culminated during the 1960s and 1970s. It is most likely caused by an increase in lake effect snow associated with colder winters from the 1950s to the late 1970s, but also may have an anthropogenic contribution in the form of heat and pollution from increased urbanization and industrialization.

(3) The winter and summer temperatures, particularly during the extreme months of January and July, have undergone some change. Cooler temperatures occurred prior to the 1930s, warmer temperatures from 1930 to the late 1950s, and cooler temperatures after the late 1950s. During the 1980s, an upswing has occurred in phase with the global increase of temperature. Values have not, however, yet approached those recorded in the 1930s.

(4) The state has experienced a long run of above normal precipitation years, interrupted, at least temporarily, by the dry year of 1987 and the early months of 1988.

The incipient upward movement in temperature and downward movement in precipitation that challenges the state in the late 1980s may or may not foretell long-lasting trends. The climatic record of the twentieth century indicates many ups and downs of climate, and an unidirectional change cannot be perceived with the possible exception of modulation. These changes affect many aspects of the state's economy including agriculture, recreation, water availability, lake levels, and energy consumption. The incapability of predicting future climate change illustrates how difficult it is to make projections of many facets of Michigan's economy.

# A BRIEF BIBLIOGRAPHY


Changery, M. J. *National Thunderstorm Frequencies for the Contiguous United States.* Asheville, N.C.: National Climatic Data Center, 1981.

Changnon, Jr., S. A. "Climate Fluctuations and Record-High Levels of Lake Michigan." *Bulletin of the American Meteorological Society* 68 (1987): 1394–1402.

Eichenlaub, V. L. *Weather and Climate of the Great Lakes Region.* Notre Dame, Ind.: University of Notre Dame Press, 1979.

______. "Lake Effect Snowfall to the Lee of the Great Lakes: Its Role in Michigan." *Bulletin of the American Meteorological Society* 51 (1970): 402–412.

______. "Some Recent Climatic Fluctuations in Michigan." *Michigan Earth Scientist* 22 (1986): 3–9.

Gale Research Co. *Climatic Normals for the* U.S. (1951–1980). Detroit: Gale, 1983.

______. *Weather of* U.S. *Cities.* Vol. II. Detroit: Gale, 1981.

Janiskee, R. L. *Comfort Climates at Grand Rapids, Michigan.* M.A. Thesis, Department of Geography, Western Michigan University, Kalamazoo, Mich., 1969.

Karl, T. R., R. G. Baldwin, and Michael G. Burgin. *Time Series of Regional Season Averages of Maximum, Minimum, and Average Temperature and Diurnal Temperature Range Across the United States:* 1901–1984. Historical Climatology Series 4–5. Asheville, N.C.: National Climatic Data Center, 1988.

Ludlum, D. M. *The American Weather Book.* Boston: Houghton Mifflin, 1982.

Lutgens, F. K., and E. J. Tarbuck. *The Atmosphere: An Introduction to Meteorology.* 4th ed. Englewood Cliffs, N.J.: Prentice Hall, 1988.

Michigan State University. *Michigan Freeze Bulletin.* Agricultural Experiment Station Research Report No. 26. East Lansing, Mich.: Michigan State University, 1965.

Michigan Weather Service. *Michigan Snow Depths.* Rev. ed., 1969. East Lansing, Mich.: 1964.

Nurnberger, F. V. *Summary Of Evaporation in Michigan.* East Lansing, Mich.: Michigan Department of Agriculture, 1976. Revised and reprinted, 1982.

Phillips, D. W., and J. A. W. McCulluch. *The Climate of the Great Lakes Basin.* Toronto: Environment Canada, Atmospheric Environment, 1972.

Ruffner, J. A. *Climates of the States.* Vol. I. Detroit: Gale Research Co., 1980.

Ruffner, J. A., and F. Bair. *The Weather Almanac.* Detroit: Gale Research Co., 1985.

Stolle, H. J. "Michigan Tornadoes: 1930–69 and 1970–80." *East Lakes Geographer* 22 (1987): 88–103.

Terjung, Werner H. "Physiological Climates of the Conterminous United States: A Bioclimatic Classification Based on Man." *Annals: Association of American Geographers* 56 (1966): 141–179.